Geographies of Solar Energy Transitions

Geographies of Solar Energy Transitions

Conflicts, controversies and cognate aspects

Edited by

Siddharth Sareen and Abigail Martin

First published in 2024 by
UCL Press
University College London
Gower Street
London WC1E 6BT

Available to download free: www.uclpress.co.uk

ISBN: 978-1-80008-727-9 (Hbk.)
ISBN: 978-1-80008-729-3 (Pbk.)
ISBN: 978-1-80008-730-9 (PDF)
ISBN: 978-1-80008-732-3 (epub)
DOI: https://doi.org/10.14324/111.9781800087309

Contents

List of figures and tables		vii
List of contributors		ix
Acknowledgements		xi

1 Cognate aspects of solar energy transitions   1
   Siddharth Sareen and Abigail Martin

2 Finance and the solar transition: project finance and tax credits
   as drivers of the US solar rollout   23
   Conor Harrison

3 Energy communities as models of social innovation,
   governance and energy transition: Spanish experiences 35
   Teresa Cuerdo-Vilches

4 Beyond power: the social situatedness of community
   solar energy systems   47
   Karla G. Cedano-Villavicencio and Ana G. Rincón-Rubio

5 Can solar energy make up for a failing grid? Solar energy
   deployment in urban and urbanising localities of the
   Global South   57
   Bérénice Girard, Alix Chaplain and Mélanie Rateau

6 Accepting idealised solar farm portrayals? Exploring
   underlying contingencies   67
   Harriet Smith, Karen Henwood and Nick Pidgeon

7 Comparative visual ethnographies of the ensconcement of solar
   photovoltaics in the urban built environment of solar cities Jaipur
   and Lisbon   81
   Siddharth Sareen

8   Community solar struggles in Portugal   97
    Abigail Martin

9   Geopolitical ecologies and gendered energy injustices for solar
    power in Ghana   115
    Ryan Stock

10 Governing solar supply chains for socio-ecological justice   125
    Dustin Mulvaney

Index   137

List of figures and tables

Figures

2.1	Typical tax equity partnership structure	30
6.1	A solar and wind farm	73
6.2	Pembroke Valero Oil refinery and Hoplass Solar Farm	75
6.3	Port Talbot Dock	77
7.1	Solar PV atop middle class housing in a Jaipur suburb	84
7.2	Solar PV integrated in Jaipur's street lighting above a shanty	84
7.3	Solar PV is ubiquitous at fuel pumps in Jaipur	86
7.4	Solar PV on an emissions testing van off the highway in Jaipur	86
7.5	Solar PV for agricultural pumpsets on the outskirts of Jaipur	87
7.6	Roofs of industries at quarrying sites on Jaipur's outskirts with solar PV	88
7.7	Solar PV and solar thermal on some of the higher Lisbon rooftops	89
7.8	Some solar PV along with a lot of unused potential on rooftops in Lisbon	89
7.9	A solar PV panel resembling roof tiles in Lisbon at a solar developer's office	90
7.10	Neatly integrated limited solar PV atop middle-class Lisbon buildings	90
7.11	A 'solar tree' at the national science museum in Lisbon's business district	91
7.12	Solar PV integrated into parking meters in Lisbon	92
8.1	Community solar: scales and owners relative to other on-grid PV systems	100
8.2	Coopérnico's vision for scaling community solar cooperatives	102
8.3	CECs and RECs	105
9.1	Location of the Kaleo Lawra solar plant in Upper West, Ghana	116

Tables

10.1 Multi-sited governance interventions currently shaping
the geography and resource assemblage for photovoltaics
production   127

List of contributors

Editors

Siddharth Sareen is Professor in Energy and Environment at the University of Stavanger and Professor II at the University of Bergen.

Abigail Martin is Research Fellow at the Science Policy Research Unit at the University of Sussex Business School.

Chapter authors (in order of appearance in the book)

Conor Harrison is Associate Professor at the University of South Carolina.

Teresa Cuerdo-Vilches is Principal Researcher at the Eduardo Torroja Institute for Construction Sciences and a professor at the European University of Madrid.

Ana Gabriela Rincón-Rubio is a researcher at the Institute of Social Research at the National Autonomous University of Mexico.

Karla G. Cedano-Villavicencio is Head of the Innovation and Futures Lab at the Institute of Renewable Energies at the National Autonomous University of Mexico.

Bérénice Girard is a researcher at the French National Research Institute for Sustainable Development.

Alix Chaplain is a temporary Teaching and Research Fellow at the Paris School of Urban Planning and Research Associate at the Center for International Studies.

Mélanie Rateau is Postdoctoral Fellow at the ESO laboratory at the University of Le Mans and Research Associate at the Technologies, Territories and Societies Laboratory.

Harriet Smith is an interdisciplinary Research Fellow in the Understanding Risk Research Group at Cardiff University.

Nick Pidgeon is Professor of Environmental Psychology and Risk and Director of the Understanding Risk Research Group at Cardiff University.

Karen Henwood is a professor in the School of Social Sciences at Cardiff University.

Ryan Stock is Assistant Professor and Director of the Illume Lab at Northern Michigan University.

Dustin Mulvaney is a professor in the Environmental Studies Department at San José State University.

Acknowledgements

We gratefully acknowledge funding from the University of Stavanger and the University of Bergen that made it possible to publish this book open access; the Research Council of Norway funding (grant 314022, Accountable Solar Energy TransitionS, ASSET project) that supported editorial work; a Peder Sather Center grant from the University of California Berkeley that funded a book workshop in Denver in March 2023; and Horizon 2020 (grant 101032239, Eurosolar4All project) and Horizon Europe (grant 101096490, RESCHOOL project) that supported editorial time for Sareen.

1
Cognate aspects of solar energy transitions
Siddharth Sareen and Abigail Martin

A short introduction to cognate aspects of solar energy transitions

Solar energy is currently the fastest growing renewable energy sector. Growth in electricity from solar energy – mostly photovoltaics (PV) – is expected to remain strong toward 2030 and beyond, with some predicting that solar energy will eventually constitute the greatest source for electricity production worldwide. In the mid-2020s, we find ourselves in the midst of a wave of solar projects of diverse configurations and contexts, from large factories to small electric generating stations, with many more projects and innovations on the horizon. To meet climate action targets, analysts show solar energy needs to expand 10–20-fold over the next two decades. How will this industry develop geographically as it scales up? To what extent will the evolution of new solar spaces of production be spatially concentrated or diffused? How are these spaces connected by similar processes and outcomes?

This book is a gathering point for a collection of enquiries that investigate the cognate aspects of solar energy deployment at a milestone-moment in the industry's development. We choose the term cognate – derived from Latin cognatus, meaning 'blood relative' – to draw attention to the diverse set of material, political, institutional and cultural developments tied to the expansion of solar industry value creation. Cognate aspects take root and grow in many diverse contexts but are fundamentally derived from the same starting place: the pursuit of solar energy transitions as a climate mitigation strategy for the electricity sector. In drawing attention to cognate aspects, this edited volume attends to the need for a more geographically-balanced view of solar energy developments as occurring not only in the economic centres of the global economy. For instance, utility-scale solar developments all over the world are rife with political and ecological conflicts between land users and sociotechnical challenges of load balancing and resource planning for grid managers. Additionally, there are examples of gender- and class-based exclusion in access to solar energy products and economic opportunity, systematic inequities in who profits from solar industry value creation and injustices tied to who bears the burdens generated by solar value creation, such as upstream manufacturing pollution and downstream end-of-life waste. Such phenomena cannot be understood as emerging in the centre and spreading to a peripheral hinterland. Rather, we propose that solar energy expansion must be understood as place-based developments that are rhizomatically tethered across the centres, peripheries and semi-peripheries of the world,[1] shaped by multiple spaces and unique histories.

In this intervention, we aim to show how the deployment of solar technology is context-specific and geographically conditioned. We argue that it exhibits certain characteristics that manifest only in particular places, while also bearing family resemblances to solar adoption in other locales and political jurisdictions. By showcasing solar deployment in under-studied regions alongside some of the contexts where it has been more famously adopted in previous decades, we bring the periphery to the fore and help problematise unfounded assumptions about technology transfer as flow from the centre into peripheries. The study of cognate aspects of solar energy deployment offers an inroad into the growing literature on energy peripheries. The concept of energy periphery situates low-carbon energy transitions within the broader landscape of uneven geographical development – itself the product of the dominant political economic system and social relations specific to that system. This scholarship draws attention to the multi-dimensional and multi-scalar nature of energy peripheralisation whereby communities remain vulnerable due to so-called 'backward' energy technologies and practices, which persist amidst increasing investments in low-carbon energy transitions. Similarly, our collective investigation underscores the potential for solar technology deployments to reinforce, rather than rectify, pre-existing injustices.

Rather than providing a survey of one specific issue taking shape across many places, we put forth a collection of essays that are purposely varied in order to help us reconsider and deepen our understanding of the nature of solar industry growth and what its role in energy transitions actually entails – where, why and for whom. This exploration takes us to a variety of solar projects and contexts across chapters. Together, the authors capture cognate aspects in diverse settings, ranging from the local to the national, including: solar support policies and investment decisions about project finance; the raw material supply chains that enable solar energy transitions; determinations about scale and ownership structures during project planning and installation; and struggles over land use, property and other resources. The book also addresses landscape changes including the built environment; integration into household and local routines; regulatory and technological grid innovations; and justice-oriented impacts on local communities and natures, including outcomes tied to environmental justice, economic equity, gender equity and more. Each chapter brings focus to one or more of these cognate aspects within a specific geographical context, while being mindful that solar energy transitions are highly globalised. The multi-sited analyses of cognates presented in these chapters also underscore how deployment practices and other dynamics occur at multiple spatial scales and in relation to multiple levels of governance of solar energy transitions.

In assembling this range of contributions, we are reminded that solar energy transitions are unfolding at a blistering pace, such that it is not yet clear how developments in one place may inform, or be part of, broader patterns of solar industry production and deployment that may become more visible in years to come. We also acknowledge that this collection is not an exhaustive presentation of cognate aspects in solar energy developments. There are notable omissions, for instance, geopolitical trade conflicts over solar PV modules and components, the growing integration of battery storage systems and corresponding hardware and software (e.g. smart electric meters) and the diverse temporalities of solar energy transitions across space compared to other sources in a given energy mix. We hope these readings demonstrate the tremendous scope to expound on cognate aspects and inspire others to add new perspectives on the aspects featured in these pages, as well as to develop new insights.

Chapters on cognate aspects

The chapters here cover a range of processes and outcomes implicated in solar energy transitions. Each contribution attends to the diverse sociopolitical and ecological contexts in which energy transitions are embedded, reflecting the highly diverse nature of solar energy transitions themselves. Some cognate aspects will be familiar to even a general audience, while others will resonate more with those who have existing knowledge and a firmer base in solar energy sociotechnical systems.

Readers will note that authors draw on meticulous and extensive empirical work, as well as deep conceptual anchoring in a range of relevant scholarship and debates. Disciplinary lenses draw from geography, sociology, politics and environmental studies, including environmental psychology. Although theoretical treatments are kept concise per chapter in order not to bury readers in details that are overly particular to narrow case contexts, we nonetheless engage with a number of important debates. For instance, some element of land and resource dispossessions linked to rapid industrial progress is present in many of the contexts examined. Conceptions of energy democracy are central to several chapters, as are questions of what kinds of techno-economic models are needed to achieve certain conceptions of democratic practice in solar energy transitions. Thus, centring cognate aspects can figure into diverse critical social science traditions.

The two chapters that follow this introduction examine how national-level policies are determining the diversity of project sizes and configurations on the ground. In Chapter 2, Conor Harrison discusses the important cognate aspect of project finance in the United States, where he argues that tax credits and the environmental, social and governance targets of big firms have driven investment in large utility-scale solar energy projects. By detailing these financing structures, the analysis reveals the limits of American supply-side industrial policy strategies and instruments and suggests that the future growth in solar manufacturing (driven primarily by recent policies like the Inflation Reduction Act) will not be as inclusive as many people hoped. Expectations for more distributed solar configurations and associated benefits will likely go unfulfilled unless additional policy and market developments occur. In Chapter 3, Teresa Cuerdo-Vilches demonstrates how national policies can encourage more inclusive growth using the case of Spain, an emergent European leader in solar energy transitions. In Spain, the rise of energy communities demonstrates multi-level alignments between European and Spanish lawmakers responding to demands for pro-poor social innovations. Implementing novel community-based configurations of solar PV and developing associated techno-economic models presents new challenges at local scales, in turn raising questions about whether and how promising community energy models can be replicated at a faster pace and broader scale.

After these two national level framings, the two subsequent chapters discuss dynamics at the hyper-local level; one in two Mexican cases and the other across comparative contexts in India, Lebanon and Benin. Both chapters highlight how solar deployments collide with local realities and everyday practices. In Chapter 4, Karla Cedano-Villavicencio and Ana Rincón-Rubio address the highly situated nature of solar energy transitions by examining social situatedness as a cognate aspect. In their treatment of community and household level cases in Mexico, they relate the diversity of energy needs to emotional topography and consider how collaborative endeavours can match solar transition strategies to contexts and aspirations for appropriate embedding in society, rather than being imposed inappropriately. Importantly, they attend to the maintenance and repair work that goes hand in hand with technological rollout in rural contexts, which often brings up unexpected challenges and adverse outcomes for intended beneficiaries. Moreover, they consider the diverse uses of solar energy not simply as electricity but as a heat source, which while well established is nonetheless underrepresented in other chapters. These notably focus on on-grid projects to a great extent, reflecting how solar energy transitions are dominantly thought of outside of remote, rural contexts in the Global South, a lacuna and bias that requires major work to address. In Chapter 5, Bérénice Girard, Alix Chaplain and Mélanie Rateau take up a cognate aspect that focuses on the heterogeneous configurations through which households and institutions access electricity, that they then mobilise in urban settings of the Global South. Here, there is explicit recognition of solar energy transitions as interwoven with a plethora of alternative energy solutions in complex urban realities, where energy needs are met through a range of both on-grid and off-grid fixes. Thinking about solar sources in conjunction with diesel generators, batteries and electric line leakages, the authors offer a richly empirical basis for centring everyday energy practices at the hyper-local scale.

Solar energy transitions layer onto existing built environments and land uses and thereby change them and everyday practices and imaginaries associated with them. Chapters 6 and 7 examine how solar PV deployments create new aesthetic landscapes, in which residents must often grapple with conflicting aesthetic sensibilities and local histories. In Chapter 6, Harriet Smith, Karen Henwood and Nick Pidgeon interrogate the visual depictions and discursive formations of solar landscapes, mainframing the cognate aspect of visions. Towards this, they draw on three solar PV interventions in South Wales that involve a real solar farm, a solar installation in an industrial cluster and a proposed solar plant envisioned as part of planned port development. In each case, there is distinctive interplay between strategic depictions and social situatedness. The authors argue that the type of work that goes into situating solar energy transitions must be understood in relation to specific societal contexts, such as local interpretations of South Wales' industrial history and memories of – as well as lived experiences in – labour exploitation. In Chapter 7, Siddharth Sareen conceptualises the integration of solar PV deployments into urban built environments as 'ensconcement', drawing on two quite different contexts: the Portuguese capital of Lisbon and the Rajasthan state capital of Jaipur in India. Examining the cognate aspect of ensconcement in both cities as sub-urban geographies reveals the bricolage of layering solar panels on to contrasting urban environments. Both cities are characterised by varied aesthetic sensibilities and affordances, demonstrating how solar energy transitions layer solar PV onto urban property. Examining how solar PV promotion trickles down from policymaking centres to the urban built environment offers insights into how solar cities may well evolve in the future, with pointers to enable planned solar built environments.

The two subsequent chapters feature a conceptual focus on social movements and on energy geopolitics and gendered injustices as cognate aspects respectively. Chapter 8 examines the pursuit of cooperative and community solar initiatives in Portugal, with a focus on the role of social movements for environmental justice and socioeconomic empowerment in Lisbon and beyond. Abigail Martin shows how social movement discourses have helped generate traction for new models of value creation and value sharing, before exploring how these groups navigate conflict with other groups (e.g. labour, utilities and for-profit start-ups) and state-based institutional innovations that have simultaneously widened market participation while dampening the potential for cooperative ownership. In Chapter 9, Ryan Stock examines energy geopolitics and gendered injustices as cognate aspects, within the empirical context of rural Ghanaian solar energy transitions. This focus brings out harrowing differences in who benefits from and who is burdened by solar energy transitions. This study reveals that the wider geopolitical relations that shape solar energy transitions in Ghana manifest systematic gendered inequities at the local scale by particular forms of project development that cut off resource access to energy poor households via land enclosures that disproportionately impact women while also exacerbating economic poverty by devastating the agrarian economy, without offering alternative employment at the solar plant. Despite gains made toward global climate goals and unlike the promises prominently promoted nationally, this grounded analysis reveals worrying tendencies to worsen local intersecting inequalities near Ghanaian solar plants.

Finally, in Chapter 10, Dustin Mulvaney centres attention on the cognate aspect of socio-ecological justice. This attends to a crucial and increasingly highlighted yet understudied part of solar energy transitions: namely, their reliance on an extensive, global and deeply problematic supply chain to produce and deploy solar modules using natural resources and labour. While there is increasing recognition of the need for ethical supply chains linked to the integration of renewable energy sources from solar into wider electricity systems based on energy flexibility and storage solutions that use batteries that require elements like lithium and cobalt, the solar PV modules themselves require numerous resources including polysilicon and land. This chapter delves into the socio-ecological injustices bound up with the processes and geographies involved in bringing about solar energy transitions, covering the full lifecycle from mining and manufacturing, through siting and deployment, to recycling and disposal. It also points to the need to consider transitions along a longer temporal scale. The result is a sobering analysis of the implications for justice as solar PV manufacturing and adoption scale unprecedented heights of volume and rapidity.

Assembling this book project involved a good run of months of reflexive engagement, including an in-person workshop on the sidelines of the American Association of Geographers annual conference in Denver in March 2023 and discussions with 'critical friends' of the project, as well as iterative editing, to fine-tune balance and fit during 2022–2024. These conversations and entanglements have lent greater clarity to each chapter and overall cohesiveness, as well as to our individual vantage points in analysing solar energy transitions. We are thus both hopeful and confident that this demonstration of the analytical purchase and generative point of departure offered by attention to cognate aspects will inspire a wider set of future efforts.

Emerging patterns and distinctions across scales: aesthetic, scalar and overarching cognate aspects

Together, these chapters draw our attention to multiple scales of energy transition. Bearing in mind that scholarly engagements with solar energy transitions problematises such spatial bounding, attention to these scales offers valuable insights into how solar energy transitions enroll different assemblages of actors, institutions and materialities. Several of the chapters traverse multiple scales – for instance moving from the hyper-local household or neighbourhood to larger urban units, or from the community and regional to the sub-national to the supra-national level – reminding us that these scales are nested, overlapping and criss-crossing in complex entangled relation. Here, we draw out three kinds of cognates – aesthetic ones that are expressed at lower spatial scales, scalar ones that are explicitly entangled with the scale of solar deployment itself and overarching ones that cut across higher spatial scales, complicated by trans-local connection and global geographies of power differentials.

Aesthestic cognates

The deployment of small-scale rooftop PV underscores the many ways in which households and neighbourhoods within larger urban and semi-urban communities ground solar energy transitions. Solar can interact with everyday practices in seemingly undramatic yet substantive ways. Although passers-by may not even see solar PV on a high city roof, small-scale systems can transform taken-for-granted rooftops and green spaces into multi-use operations for energy production alongside clothes-drying, kite-flying, sunset-watching and garden-tending. These and other interjections of small-scale energy production into quotidian life can spur changes in household labour and energy consumption or changes in one's perceptions of (and interactions with) the built and natural environments. The construction of larger utility-scale solar parks can spur more dramatic shifts in land use or loss of certain human and non-human land users – such as the loss of sacred trees, animals, pollinators, fungus and other beings – transforming an existing landscape in ways that alter animal migrations or roads and walkways for human access to nearby forests and other resources for personal use. Such changes can interrupt expectations of aesthetic experience, be it the view of passers-by, or local narratives about the land that reference the physical landscape. In this way, attention to aesthetic cognates pulls focus on to the hyper-local experiences of solar and underscores the range of potential impactful interactions that may shift behaviours and intertwine with broader elements of cultural and social change.

As Smith et al. aptly demonstrate in Chapter 6, aesthetics matter here. The images created by solar – real or proposed – can act as repositories of historic and cultural meaning about a particular landscape. Residents who are most impacted by the solar power plant may have diverse understandings about how solar energy brings change to their day-to-day life and how solar energy figures into their recent history and impending future. In contrast, solar plant owners may well never see the everyday impacts of globalised finance that deeply inflect energy infrastructure ownership and benefit flows. Interests rarely align across remote owners and local residents. Tensions between project developers' and owners' profit-maximisation motives and locals' values can play out in aesthetic conflicts about how solar 'looks' in a certain place, as well as in public debates about whose vision of solar energy is most appropriate.

Thus, examining how aesthetic experiences play out at the hyper-local level brings fresh perspectives to the literature on energy *visions and imaginaries*, underscoring the diversity of place-based meanings imputed to and held by those impacted by solar energy transitions but not typically involved in decision-making and thus not easily discerned at higher levels of analysis. As a cognate aspect, aesthetics draw our attention to the fundamentally visual and discursive constructions of solar in a variety of media, from illustrations to political speeches, advertising campaigns and the promotional materials generated for corporate shareholders or resident welfare associations. Attention to how solar is envisioned helps highlight the different framings actors use to explain what solar energy involves, whom it will benefit and why any undesirable impacts should be tolerated. For instance, planning for large solar plants often entails a slew of ephemera – such as brochures, social media posts, accounting sheets – designed to shore up performative engagement activities such as regulatory consultations with local stakeholders or corporate public engagement events. These materials often work to rehearse key claims about the co-benefits of solar projects in terms of community welfare achieved through investments in public parks and local schools, educational scholarships and community grants.

Hyper-local investigations into the aesthetic cognates of solar may also shed light on the disconnect between community members and solar project developers in terms of what is required to obtain a formal or informal social license to operate. Those who do not inhabit a landscape may remain unaware of alternative conceptions of value and risk that are rooted in local lived experiences. Often, industry visions of utility solar go unchallenged and materialise in ways that change societal rhythms without local consent, as people adapt their everyday practices to a new landscape that is less hospitable to local tastes, values and routines. A neighbour of a large field of photovoltaic panels who once saw cattle grazing against the sunrise may now shield their face from bright reflective glares. These and other experiences can undermine or contribute to local satisfaction about advancing climate change mitigation.

In addition, attention to aesthetic cognates in urban and peri-urban environments may draw attention to the casual creep of solar PV onto the built environment, without much consideration for how it impacts everyday experiences. In urban and peri-urban contexts, many – perhaps most – solar installations are grafted on to seemingly passive, empty roofs and parking lots overnight, whereas behind closed doors a family or community group has quietly organised itself into a prosumer to capture the benefits of reduced electricity bills and elsewhere a solar installer has configured inverters, modules and backing sheets onto old buildings, new battery units and power grids.

As Sareen shows in Chapter 7, the act of this layering atop existing infrastructure, and the social relationships underlying such layers, render the aesthetic cognates of solar such as ensconcement phenomenologically important to explore. A field stops growing vegetation to host panels and inverters, or an agrivoltaic arrangement requires new planting and harvesting methods, inputs and technologies. Mowers and other equipment replace a few sheep in a former pasture. A roof once decorated with damp clothes is now dressed by a tangle of wires, traversing attic spaces that hide inverters, batteries and monitors from the rain. At stake here is not only visual change, but concomitant everyday alterations of the local sociotechnical fabric.

Ensconcement can also play a significant role in advancing or eroding local legitimacy of solar energy deployment, in both rural and urban contexts where solar installations compete with many existing uses of limited space and desires linked to future developments. In urban spaces, photovoltaics may be legally permitted but deeply unpopular. For instance, adding photovoltaics atop the culturally-valued ceramic tiles that contribute to the splendid colours in Portugal's capital city may violate the rule of a planning or heritage authority. In addition, some Lisbon residents indicate that panels reflecting unwanted sunlight into other buildings or blocking cherished views could generate interpersonal conflict. Similarly, in Brooklyn, New York, some neighbours frown up at solar arrays mounted onto historic Bed-Stuy brownstones, which may disrupt a beloved skyline or confirm suspicions about unwanted demographic change.

Elsewhere, cities wrapped in solar panels may serve as symbols of progress and a hopeful energy future. Indeed, more acceptable aesthetics are often linked to popularisation campaigns and policy pushes, such as the New European Bauhaus. A large solar plant taking over peri-urban land may be more welcomed by the suburban neighbourhood if it is accompanied by public investment in a local community centre to raise awareness about energy transitions, or by small solar PV fit on to local streetlights to lower municipal spending as part of the same project. These and other vignettes suggest that public institutions can play important roles in creating positive models for ensconcement, for instance by developing arrays on public school buildings, town halls and elsewhere, in ways that take aesthetics into account and hold space for community debate about the implications of aesthetic choices. If small and large enterprises follow suit, the rollout of solar PV could achieve ubiquity without backlash to many a changing landscape. In Europe, a large part of what may or may not happen and how soon is likely to be determined by policies for energy communities.

Aesthetic cognates thus bring together the visions, ensconcement and social relations that play out at hyper-local, urban and peri-urban scales, underscoring how these cognate aspects hold enormous significance for the ways in which solar energy is sculpted onto landscapes and for which actors benefit or perceive harm from emergent configurations. Solar PV is the fastest expanding energy source in the world and its form and effects at the hyper-local scale will continue to grow in importance. Even as low-hanging fruits get used up and conflicts escalate with greater competition for limited space and land to install solar capacity, respectful and aesthetic ways of spatialising and situating solar PV also continue to multiply and will undoubtedly proliferate. Societal visions should be attuned to the need for contextually appropriate manifestations, for their efficient ensconcement into built environments and everyday practices and for multiple forms of usage and sharing through solar energy communities.

Scalar cognates

The scalar cognates of solar energy transitions spotlight the wide range of social and material configurations in solar energy transitions, from small residential rooftop systems up to utility-scale power generation installations. Utility-scale solar has traditionally been the main source of growth in PV installations and continues to be the least costly option for new electricity generation in most countries in the mid-2020s, despite the cooling effect of rising commodity pricing on investors. However, distributed solar PV, such as rooftop solar on buildings, has also experienced considerable growth due to higher retail electricity prices and various forms of policy support. As Martin illustrates in Chapter 8, different configurations entail different scales and owners, with community-scale configurations bringing new actors into the landscape of both centralised solar farms and distributed rooftop systems.

Attention to scalar cognates also illuminates the challenging politics of utility-dominated power sectors. Many industry analysts believe that unless small-scale, distributed solar plants are adopted by diverse users – from pensioners to young families, from retail shop owners to municipalities – the benefits of solar energy transitions will remain limited to first-movers and the PV industry will fall short of its market potential. However, utility companies actively and successfully lobby to weaken or remove subsidies for small-scale distributed solar, which has generally flourished under attractive tariffs. Fears about demand flight and situations whereby high penetration of solar to the grid from large-scale plants sends prices plummeting during peak hours limit value creation for small owners who lack the means to survive such volatility.

Community-scale solar projects have also struggled under utility-dominated regulatory politics, although European advocates have succeeded in pushing lawmakers to pass laws that enable more distributed, collectively owned and democratically governed forms of energy production, often termed solar (or renewable) energy communities. Here, the concept of communities can – but does not necessarily – refer to socially relatively tight-knit collectives, typically co-located in a neighbourhood or settlement. Rather, solar energy communities can feature considerable internal diversity along intersectional lines of class, caste, ethnicity, gender, race and other forms of social status and capital, as has been much discussed in critical energy social science. Still, the legal basis for energy communities features considerable variation; in many places they remain illegal or highly disincentivised by regulations biased towards incumbent utility companies and lawmakers reluctant to disrupt top-down, centralised energy infrastructure and institutions.

Cuerdo-Vilches shows in Chapter 3 and Martin in Chapter 8, that this situation is not lost on governments. Spain, Portugal and other European countries have enacted enabling laws and regulatory changes for solar energy communities. Participants in such developments are often motivated by the opportunity to establish shared solar energy production systems, anchored in the community and governed by local collaborators interested in local benefits and a more democratic electricity sector. However, as Martin details, this space is also diversifying as for-profit actors move into these markets in an attempt to establish commercial cognates of entities originally envisioned as cooperatives.

Thus, examining scalar cognates attends to the considerable variation in ownership and technoeconomic models for community solar projects, which reflects the diversity of socioeconomic needs, value-creating desires, energy rhythms, electricity portfolios and end-user capabilities that energy communities may involve. Further, energy communities may not necessarily correspond to the spatial and social characteristics of other neighbourhood-based communities. One can be part of an energy community by owning shares and exercising a decision-making stake and/or by participating in the actual generation, consumption and sale of solar power. Depending on legal definitions in a given context, one may live within a radius of a few kilometres of other energy community users in a rural context, yet be connected by the same grid and consume virtually from one larger solar plant. Or one may live in the same cooperative housing and be involved in deploying multiple energy flexibility solutions based on shared rooftop PV plants and batteries wired to neighbours in the same and nearby buildings and perhaps to proximate shared vehicle charging infrastructure.

As such, the study of scalar cognates reinforces a key insight from the above discussion of aesthetic cognates, namely that there is no one ideal solar energy transition model or scale and that diversity can bring more reward than risk. If solar energy transitions were to localise a great deal of energy generation in and around the points of end consumption, with the intent of increasing ordinary people's ownership of and degree of control over energy infrastructure, there is some potential for local power relations to warp the distribution of benefits to households within a community. The rationale is that those in privileged positions would exercise their superior status to accumulate greater wealth than those in marginalised positions, thus reproducing social inequality at the community scale. Yet this diversity need not inherently create conflict. Rather, it also constitutes a potential asset for those interested in diversifying PV configurations to enable different arrangements to work for cooperating entites. Variety at the community scale allows for a number of diverse solar PV forms to proliferate and for different arrangements to work across many cooperating entities within a given community.

This point is brought to bear richly upon the Mexican case studies that Cedano-Villavicencio and Rincón-Rubio analyse in Chapter 4, where they argue for consideration of social situatedness as a cognate aspect. The subjects they engage with in these communities have specific, situated aspirations related to solar energy: they want it in order to power particular activities as part of their enterprise, or to open up some possibility. Research on energy access has shown that access can open up many different doors, just as it can also create dependencies that can then go unfulfilled if equipment fails, maintenance problems emerge and long-term availability issues go unaddressed. Emphasis on social situatedness thus brings other issues into consideration, such as the actual practices bound up with solar energy within a societal context at the community scale. Is there a way to configure multiple needs so as to put the peculiar productive rhythms of solar energy to work across multiple needs and competencies in a given context? Can someone draw upon power during the day (such as residents of a home for the elderly, or a local grocery shop that runs refrigerators and freezers, or a carpenter with power tools), while others use it during the late afternoon hours (a family with young children cooking a meal and someone running a laundry load) and yet others use their technical savvy to hook up a battery to the panels to charge their vehicle or to sell it to the grid when prices are high? Such synergies require ties of cooperation and enabling regulatory mechanisms.

Examining scalar cognates in the context of community solar helps to frame the social position of these projects, especially the conflicts and tensions between different community members over how energy decisions are made. For instance, patriarchal households and organisations where men dominate energy decision-making often sideline or ignore the distinct needs, rhythms and practices of other genders, creating the conditions for energy vulnerabilities to persist despite the availability of solutions or needed resources. There may also be intergenerational tensions between older people who have 'grown up' alongside currently established energy practices and younger people who are presumed to lack know-how in energy-related matters. Of course, such intergenerational expectations and tensions vary across societies. Still, ignoring the age- or gender-related needs of certain groups may lead to blind spots in the governance of solar energy transitions.

Investigating scalar cognates also draws attention to cross-sectoral developments, such as the presence of particular occupations and land use patterns – e.g. farming crops that thrive in the shade – that facilitate novel pairings such as agrivoltaics, which combines agriculture with solar plants to maximise value generation from farmland in many geographical contexts. The presence of a big power-hungry industry in its proximity, that can serve as an offtaker to absorb any extra local solar energy production, can be a boon to install solar plants at the community scale. Recognising and cataloguing these priorities for solar energy transitions in social settings is a key step towards pushing for empowering legislative and regulatory changes that can make solar plants relevant for communities.

Scalar cognates can also highlight the struggle to create desired value from solar energy configurations for certain groups vis-à-vis others. For instance, a company may prize reliable electricity access from the grid far over a community solar plant. But it may decide to put up solar PV on its warehouses and reduce its demand from the grid while making better use of existing assets (roof space) and capturing a large share of benefits from the new solar PV value stream (lowering costs). While the developer may be happy with this arrangement, the community may have preferred a local offtaker, but it does free up space on the existing electricity grid for them to stream their generation and provides them with a helpful point of fact that may be used to argue against the utility in case the latter takes a protectionist stance and lobbies its regulator not to enable local community-scale energy production.

In Chapter 5, Girard, Chaplain and Rateau make this point about diverse options and sources of energy by articulating heterogeneous electricity configurations as a cognate aspect. Heterogeneous electricity configurations refer to the amalgam of different energy sources that come together to fulfil energy needs in a given context. In the urban Global South contexts that are the empirical setting of Chapter 5, there is considerably more diversity in these sources than in say a rich northwestern European country like Norway that powers household energy services primarily with electricity. Yet, in many parts of the world solar energy is entering contexts where fossil fuels remain an important, even dominant, part of the energy mix to meet people's basic needs and the indulgences of the elite. Solar energy thus enters contexts where it not only has to displace incumbent energy sources, but intertwined energy practices, which are bound up in complex sociotechnical relations such as other infrastructure beyond energy, social norms and simply habits that people have not reflected deeply upon or find easy to shift away from. Rather than try to change everything all at once, a workable strategy for the solar industry is to fit specific needs and substitute other fuels there first, to gain a foothold and become a familiar part of energy systems within communities, before its use can be expanded more significantly. In some cases it may also work best as a complement to other sources, given existing established infrastructures and policies. While this can evolve over time, within a short time horizon where solar energy continues to expand rapidly, heterogeneous electricity configurations are an empirically highly applicable and logical way to consider solar energy transitions.

Other dimensions brought forth by attending to scalar cognates are the ways in which solar projects across different countries may be linked together by lawmakers acting in concert to facilitate community solar projects, as well as networks of advocates for a particular configuration of solar technologies and governing institutions. In Chapter 8, Martin examines solar cooperatives and highlights the role of European social movements in formal policy arenas and coalition-building to advocate for alternatives to utility-scale arrangements and individual self-consumption systems. Such community solar advocates leverage claims about the need for more just and democratic solar energy development, linking community solar to numerous socio-ecological issues to broaden political support. This reflects how solar energy rollouts intersect with long-running social and environmental injustices, such as racial- and class-based divides in access to solar energy production and consumption, precarious labour in rooftop installations and rising energy and housing prices due to neoliberal markets, despite the mitigating role of cooperatives. A salient point is that, without attending to the social situatedness of solar, efforts to address social justice by passing off-the-shelf legislation can frustrate solar deployment.

If social situatedness directs attention to the community scale, heterogeneous electricity configurations speak to the realities of energy usage at this scale. Further, the cognate aspect of social movements serves to remind us of the values that are important to uphold in terms of the community impact of solar energy transitions – fair jobs, affordable clean energy access for all and safeguards against neoliberalisation of other related essentials such as housing due to the advance of solar PV. These challenges are not inherent to solar energy itself, which is a far less troubling energy source than most others. Yet, since solar energy transitions enter into implementation in community contexts all too often characterised by existing inequalities, it is crucial to bear in mind that these inequalities should not be exacerbated.

Overarching cognates

While some impacts of solar energy transitions are felt proximately and their fit-to-context requires recognising and addressing various sociotechnical factors and power dynamics at the hyper-local and community scale, a great deal of what shapes solar PV takes place at higher scales like the national and global. This includes translocal connection, such as in the supply chains, labour geographies and eventual waste streams related to solar modules. But it also includes that invisible yet arguably most powerful of structuring mechanisms – money – as it is the availability of solar finance that allows projects to be licensed and installed. Energy has a preeminent role in politics. This is linked to supranational concerns of energy security; national emphasis on energy access, affordability and reliability; and regional politics of who benefits from and bears the burden of energy supply within the national picture. Thus, energy geopolitics have an overarching effect on the patterns that solar energy transitions display across countries.

As Harrison shows so convincingly for the USA in Chapter 2, modalities of energy finance comprise a key cognate aspect of solar energy transitions. Subsidies, tax credits and other forms of direct or indirect support that solar developers and financiers are able to avail themselves of through complex maneuvering impact where solar plants are installed, under whose ownership, with what generation capacity and with any supplementary local initiatives. Importantly, these financial mechanisms are not evident to ordinary people and only a few (such as feed-in tariffs for household rooftop solar PV) are typically accessible to entities other than large (and often multinational) corporations and/or banks. Even the minimal small-scale solar support schemes have been phased out in most places in recognition of the competitive cost of solar PV relative to other energy sources and if anything, households often fare relatively poorly in solar finance configurations by having to take low tariffs for their sale to the grid. Yet, where states have continued to put incentives in place – albeit more complex ones than high solar feed-in tariffs, which were especially popular in the 2010s – is in large solar PV parks. These help governments meet their ambitious climate targets and reduce bureaucratic and technocratic demands, by being signed off to a few large players who install large PV capacity.

The downsides that go with this approach – which reproduces old logics of a top-down, spatially concentrated nodal energy sector while forgoing the modular nature and beneficial distributed possibilities of solar energy transitions – include a lack of democratic ownership, persistent control over energy by remote, large and often private corporations and the accumulation of capital to a few wealthy and influential entities rather than to a wider range of residents in the areas neighbouring solar deployment. Crucially, large-scale solar parks are also reliant on massive electricity transmission grids, which are often expanded linked to these developments – an expensive, carbon-intensive and slow process with the costs often borne by the state. This means a loss of the energy efficiency associated with spatialising solar PV close to pockets of energy consumption, reducing the need for transmission and instead investing in strengthening local distribution grid infrastructure, while also minimising the losses entailed in transmitting electricity over large distances. At its best, energy finance can accelerate solar energy transitions in context-sensitive ways at multiple scales. Based on current evidence, in practice it functions rather sub-optimally. That said, solar energy has become firmly established as the fastest growing energy source worldwide in the 2020s, beating the expectations of most industry analysts a few years ago, so its enabling and transformative role is nothing to be scoffed at.

What this rapid flow of money has led to, however, includes forms of violence that Stock vividly illustrates in Chapter 9, when addressing the impacts of solar PV development in Ghana on proximate residents and also its gendered injustices in these impacted communities. Those directing global finance flows and with an interest in promoting energy transitions have little time or appreciation of the complex webs of social relations that solar plants intervene in. If they have visited these sites, it is most likely for a short period, to gain a technical overview before signing a project agreement, or to inaugurate an installed plant. Their attention is captivated by financial policies, clearance processes and profit margins, which are getting squeezed in solar projects in many countries. Yet, for the women living in villages where they cross fields to access vital natural resources from the nearby forest, having those routes blocked by land enclosure for solar plants means spending extra hours every day for the same tasks that took less labour in their busy lives. For the hopeful residents who do not see locally sited solar plants translate into local jobs, but rather only bring in temporary work – primarily done by labourers brought from elsewhere – there is little joy in knowing that this development serves to advance climate mitigation. Enclosure takes over land near them, territorialising landscapes to extract value for remote firms, both in the form of profits from solar electricity production and as electricity itself, which is transmitted away from the local community to power-hungry consumers elsewhere.

As countries and companies work to achieve their climate targets, carbon metrics play a reductive role where climate mitigation trumps other forms of value, such as local empowerment and development that solar PV could well contribute with greater care on the part of funders in setting their terms of reference. The cognate aspect of energy geopolitics and gendered injustices serves to highlight this link between the global and the local, shedding light on how, in the absence of adequate foresight and consideration, global partnerships and national commitments open up localities to interventions that adversely impact residents. All the responsibility for this does not rest at any one scale, but rather the problem stems from the fact that scales are nested and that a different calculus operates in decision-making at the global and national levels on the one hand and at the local and hyper-local level on the other. One can hardly attribute gendered injustices in a Ghanaian community or household to remote solar energy project financiers or local installers alone; rather, the injustice arises from a combination of unaccountable interventions and deeply embedded traditions (such as gendered roles in the division of domestic labour) that make land enclosure especially problematic for those whose practices rely on long-running access. This cognate aspect highlights a failure to acknowledge and address these adverse effects as an essential part of governing solar energy transitions that governments have abdicated.

This brings us squarely to the last cognate aspect addressed in this book, namely socio-ecological justice. In Chapter 10, Mulvaney takes us through the supply chain of solar energy transitions, from mining and manufacture to siting and deployment and thereafter to recycling and disposal. There are known problems in the mining and manufacture of solar PV, related to labour rights abuse and a lack of adequate transparency (which itself is problematic) and by now considerable evidence of dispossession of marginalised groups from land taken over by large solar parks in siting. However, justice issues related to recycling and disposal are going to explode in years to come, unless urgent action is taken to create and enforce standards related to reclaiming materials from end-of-life solar equipment. As unprecedented volumes of solar modules are deployed, competition over supply chains, manufacturing facilities and sites heats up. This can be a race to the bottom, or with ethical standards championed by frontrunners, a hopeful pathway towards solar energy transitions that embody the logic of a circular economy. This is not consistent with the cheapest deployment possible and nor is manufacture with ethical labour conditions. Rather, there is a need for valuing justice in the whole lifecycle of solar technologies in a manner that impacts the market mechanisms through which solar plants secure funding and get built. This also points to the need to unpack temporality as another cognate aspect of solar energy transitions in future work.

Another concern that focusing on socio-ecological justice raises is that solar energy transitions require multiple competencies to govern. They involve material expertise, a rapidly evolving and global network of suppliers, financiers, developers and installers and those in charge of repair and maintenance, each with their core technical expertise and various forms of social and financial capital. In due course, an increasing number of players will also take up the task of recycling and disposal, more so than today. Getting governance right across all these diverse fields of operation takes a hands-on understanding of the challenges these actors face in implementing all of their tasks at the breakneck speed of sectoral development with heightened competition. Without such an understanding and requisite governance measures that help actually secure local interests, these various experts will hardly fulfil a responsibility towards others that governments do not make a sincere effort to safeguard and prioritise. If everyone feels hard-pressed in a sector whose success is an essential piece of the climate mitigation puzzle, then the ones most likely to get squeezed are the ones least able to do anything about it. As we have seen above, this lack of socio-ecological justice is acutely felt proximate to solar deployment sites, where solar plants change local contexts for the worse without giving much in return. Yet, as ample evidence also suggests, it does not have to be this way.

Hence, overall, we see that energy finance, energy geopolitics and gendered injustices and socio-ecological justice are of a piece. They all relate to overarching ways in which solar energy transitions take place and can be improved. There is no dearth of solutions and even willing takers, but there are many splintered actors each directed by quite distinct interests ranging from the pressures of profit-maximisation to the desperation of being dispossessed of one's already sparse resource base. While it is natural to have greater sympathy for those at the losing end of the bargain – the veritable 'wretched of the earth' – joined-up governance also requires an appreciation of the perspectives of other actors who together constitute the assemblage of solar energy transitions. Some of these perspectives need to be reshaped with carrots and sticks, others to be worked through collectively by putting to work that oldest of techniques, deliberative consensus, where concerns are voiced and heard, then resolved in a spirit of mutual aid and solidarity.

Centring cognate aspects: pluralist solar energy transitions research

In the final section of this introduction, we are keen to reflect upon the guiding intent of this book and the overall contribution that we hope it will make. Solar energy is the world's largest growing source. Its growth in the 2020s and beyond will have effects far beyond electricity generation. The chapters to follow focus on cognate challenges, controversies and conflicts of solar rollouts in diverse geographies of energy transition. As we have advertised to whet your appetite, these cognate aspects relate to formulating new place-specific solar energy visions and strategies, financing specific deployment scales, expanding or replacing electricity infrastructure, accessing land and resolving conflicts surrounding competing land uses. They also extend to incorporating energy storage and charging technologies, adopting flexible energy production and consumption relationships, displacing fossil fuel energy production with renewables, enabling new energy ownership models and addressing the many environmental and social injustices across the value chain of solar expansion, including upstream extractivism and downstream waste. Across these thrusts, this editorial introduction has centred how solar energy governance (both state-based regulations and more market-driven modes of governance) is evolving – and indeed must evolve – to address these cognate aspects in diverse empirical settings.

The chapters conjointly aim to convey a realistic sense of the vast range of cognate aspects of solar energy transitions and necessarily encompass multiple related sectors. These aspects are typically tangential to the main focus of scholarship on the governance of solar energy transitions. By placing them front and centre, the book aims to direct attention to the many wider changes in society that are necessitated due to the emergence of solar energy as the greatest growing source to power societies during the 2020s and beyond (estimated to overtake coal by 2027, according to the International Energy Agency). The scope covers a diverse set of case studies that illustrate various cognate aspects. Our introduction has aimed to offer a synthesised overview of cognate aspects of solar energy transitions. While not aimed to be comprehensive, we hope and trust that this range will allow the book to span a necessarily dramatic scope as an early thematic contribution.

The collective effort behind this book is to bring to light cognate aspects of solar energy transitions in an engaging manner that reaches a wide audience, including both energy sector specialists and non-specialist readers. We are also keen to engage those with an interest in other sectors that intersect with solar energy, such as land, finance, transport and resource extraction. We expect and intend for book chapters to be used by readers to sample case studies of solar energy transition as well as to inform conceptual framings of cognate aspects of solar energy transitions.

In closing, we make two claims about intent, two claims about achievement and express one wish for readers to fulfil. We claim that this book offers diverse case studies on the range of cognate aspects typically linked to solar energy transitions. We also claim that it advances understanding of the politics, policies and processes of low-carbon energy transitions, with special reference to solar energy. Coming to what we hope to have achieved: we illustrate how solar energy transitions intersect with changes in other sectors like land, finance, manufacturing and mining. We moreover contextualise solar energy transitions in a range of physical, political and economic geographies. Finally, we wish for our readers to be inspired by this form of collective engagement with and in solar energy transitions and hope that we are able to reach a wide audience of scholars, students and professionals.

Notes

1   Various scholars inspire the rhizome metaphor, but we are primarily drawing from Gilles Deleuze and Félix Guattari in *A Thousand Plateaus: Capitalism and Schizophrenia* (Minneapolis: University of Minnesota Press, 1987), in which they make concrete references to plant growth. Shoots and stems grow outward from nodal underground root networks to challenge then-traditional conceptions of imperial racist practice. Other scholars have used the metaphor of rhizome to describe social relations and power that do not operate according to hierarchy and the importance of rhizomatic relations for generating the kind of political power that could enable real counter-hegemonic possibilities, such as Michael Hardt and Antonio Negri in *Empire* (Cambridge, MA: Harvard University Press, 2000) and Peter Evans in *Is an Alternative Globalization Possible?* (*Politics & Society*, 36 (2), 2008, pp. 271–305).

2
Finance and the solar transition: project finance and tax credits as drivers of the US solar rollout

Conor Harrison

Introduction

This chapter considers the rise of solar energy in the United States of America (hereafter US) from the perspective of the financial industry. I focus on one particular aspect of solar's growing prominence in the US – the prevalence of tax equity investors – in order to demonstrate the ways in which using the lens of finance can open up new perspectives on how, where and why solar is being deployed, as well as the crucial question of who benefits. By focusing on finance, I highlight how the quest for and ability to realise profits and not necessarily the falling price of solar powered electricity generation, has been the key driver of solar development in the US. The recently passed Inflation Reduction Act (IRA) seems set to accelerate the solar industry and deepen the industry's dependence on tax investors to underpin new solar development.

The signing of the IRA in August 2022 has been widely celebrated as a major step towards the US transitioning away from fossil fuels and towards renewable energy. The bill sets aside $370 billion for supporting low carbon forms of energy, making it the largest sum devoted to clean energy in US history. The passage of the bill came after years of political wrangling and ultimately at a time when it appeared that the US (and indeed, global) renewable energy industry was heading into a post-subsidy era (Christophers, 2022). Yet the IRA serves to prop up – and likely supercharge – a solar industry that was already booming. While electricity generation from solar comprised 5 per cent of electricity generated in 2022, nearly 50 per cent of the new electricity generation capacity added in 2022 came from solar. Indeed, any number of projections show that a majority portion of future generation capacity addition in the US will be from solar PV.

Yet, there are important questions about the nature of the US solar boom. As Baker (2022) elucidates, the global renewable energy transition is being shaped by two competing objectives. The first is the potential of renewable energy as a force for community empowerment and socio-economic change. This objective is best captured by the various discourses around energy democracy, wherein renewable energy holds the potential to unwind decades of centralised control by investor-owned electric utilities and related fossil fuel interests and facilitate a shift in ownership towards more localised and/or public control. The second – and competing – objective is to bring renewable energy more firmly into the realm of the private sector and to make renewable energy attractive as a long-term revenue stream for investors (Baker, 2022).

This chapter draws on fieldwork undertaken between 2018 and 2023 that includes more than 50 interviews I conducted with financial actors involved in the US energy transition, as well as participant observation at numerous energy finance conferences. I use this fieldwork to outline a financial feature of the ongoing solar boom in the US that remains at the forefront of the US solar industry even with the passage of the IRA: tax credits and so-called 'tax equity' investors. Tax incentives and primarily the federal investment tax credit have been the primary method of subsidising solar energy development in the US since 2006 (Knuth, 2021). But what is notable about the emergence of solar as one of the premier energy investment vehicles in the US is that it is layered upon a longer history of financialised investment in the US electricity system. Section 2 discusses this longer history of the finance–state–electricity intersection in the US. In doing so, I frame the recent developments around tax equity partnerships in the longer history of electricity capital in the US (Luke and Huber, 2022).

Section 3 focuses on how solar farms are developed and as such reveals how solar has emerged as an asset class ripe for investment in the US – what Kennedy and Stock (2022) refer to as solar as an accumulation strategy. Finally, as I describe in Section 4, the use of tax credits as a financial incentive has had a shaping effect not only on the financial patterns of solar development, financing and ownership, but also on the types of installations that have become prevalent. While the passage of the IRA provides some changes to how the tax credit can be monetised by investors, the IRA ensures that the tax credit – and private ownership – remains the centrepiece of solar development in the US (Christophers, 2023).

In general, the chapter demonstrates the value of examining questions of finance as a cognate aspect of solar. My focus on finance in this chapter is shaped by the work of Christophers (2022), who argues that the pace and location of the transition to renewables has been – and likely will continue to be – driven by matters of investment, rather than simply the declining cost of generating renewable energy. And investment, Christophers argues, is not about price. Rather, investment is about the potential to realise *profits*. What is more, the very nature of solar (and wind) makes questions of finance central, even compared to other forms of energy generation. Because solar and wind require no fuel and relatively minimal oversight and maintenance over their lifetime, nearly all the costs of constructing these plants are in the initial capital cost. As such, differences in the financial aspects of renewable energy facilities can spell the difference between a project that moves forward and one that is never built. So, while the falling cost of solar has made it competitive or even cheaper than most other generation technologies, the ability to make profitable investments in solar – something largely dependent upon finance – is behind the recent rise of solar energy in the US. Tax credits have been, and will continue to be, key to this profitability.

Finance and the US electricity system: a brief history

Scholars have recently engaged in extended debates over the temporal and spatial nature of what is commonly termed 'financialisation'. For many, financialisation describes a wider transformation of the economy and society towards finance-led economic systems, most notably in the prominence of accumulation through financial channels rather than via commodity trade and production (French et al., 2011; Krippner, 2005). At the same time, financialisation has been used to describe the growing power of financial values and technologies on corporations, individuals, and households (Christophers, 2015; Langley, 2018). Yet the brief history of electricity through the lens of finance I describe in this section demonstrates that the industry has long been 'financialised', in large part because of the necessity of making large investments in fixed infrastructure that are only able to be paid back over very long periods of time (Harvey, 2006). This material reality of electricity investments has given rise to what Luke and Huber (2022: p.1699) describe as 'electricity capital' – 'the nexus of state, regulatory, and financial relationships that shape private accumulation through electricity provision'. As I show in this section, a key role of the state in this relationship is ensuring the necessary preconditions for value capture by the private sector – via funding technological development, or, more frequently, setting favourable regulatory and/or financial incentive structures.

The intersection of electricity and finance in the US can be traced back to its origins in Edison's Menlo Park laboratories – a location where much of the work was funded by venture capitalist (using contemporary language) J.P. Morgan. Not surprisingly, Edison's original central station electricity generation on Pearl Street in Lower Manhattan was located just blocks from Morgan's offices on Wall Street (Hughes, 1983). Decades later and in the midst of cut-throat competition, electric utility entrepreneurs like Samuel Insull sought out regulation by state governments in large part to solve their financing problems. In this case, state regulation was paired with a monopoly service territory, an arrangement that moved the electricity industry from a risky fringe investment to one with nearly guaranteed long-term revenue certainty (Harrison, 2013; Hausman and Neufeld, 2002). Yet, with the certainty of revenue came the potential for financial engineering. Throughout the 1920s, electric utility companies established a complex system of layered holding companies that grossly inflated stock values, and that ultimately was a major contributor to the Wall Street crash of 1929 (Hughes, 1983).

While state-level regulation served as a method of 'de-risking' investment in electric utilities, the financial misdeeds of the 1920s brought on an expanded role for the government in the US electricity system during the 1930s. Franklin D. Roosevelt's agenda for the electricity sector was largely defined by a pushback against the excesses of private utility ownership. In addition to passing laws that limited the use of holding companies, New Deal programmes created the Tennessee Valley Authority and the Rural Electrification Administration, both of which provided a new way to overcome the financial challenges posed by the electricity system. These initiatives used federal ownership as well as low-cost loans to cooperatively owned utilities (over investor-owned ones) to finance electricity infrastructure (Brown, 1980; Spinak, 2020).

However, after public power peaked in the 1950s, the rest of the century saw the gradual clawback of power by investor-owned utilities in the US. Along the way, the state continued to de-risk the electric utility space for investors via favourable regulation around matters of taxation and depreciation (Hyman et al., 2005), as well as through federal programmes to support the privatised rollout of nuclear power (Mahaffey, 2010). By the 1990s, the deregulation fervour that had captured most other regulated industries in the US (e.g. trucking, telecommunications) swept into the electricity sector. While de-risking the electricity sector via state regulation had created a stable investment, a new group of investors and electricity sector actors saw the potential to realise greater profits in the staid regulated utility business. For example, Enron viewed the introduction of markets into the sector as providing opportunities to trade electricity and use new modes of financial engineering to supercharge profits (Froud et al., 2004). Major investment banks also got on board, particularly as it became apparent that deregulating the US electricity system would come with a power plant construction boom, as independent power producers (IPPs) sought to join competitive electricity markets (Isser, 2015).

As portions of the US moved towards deregulated electricity markets, the role of the state shifted from regulator of a monopoly utility to the facilitator of competition. However, periodic crises in portions of the deregulated electricity grid (e.g. California power shortages in the late 1990s, the Northeastern US blackout of 2003, and the Texas Freeze of 2021) have typically been followed by calls for stronger state regulation to prop up investments in the electricity system.

It is in this financial and regulatory context that the transition to renewable energy has begun. To date, 74 per cent of non-hydro renewable energy operating in the US has been built by IPPs rather than by investor-owned utilities (Huber and Stafford, 2022). As such, renewable energy developers in the US have had to shoulder more of the risk of plant construction without the guaranteed revenue (and near certain profits) that had historically bolstered power plant development. To offset this increased risk, the federal government has sought other ways to de-risk investment in renewables in ways that facilitated private sector profits. While there have been a variety of policies that have aided in renewable energy development (chiefly state level renewable portfolio standards, and standard off-take agreements for qualifying facilities under the Public Utility Regulatory Policies Act of 1978), the primary financial mechanism has come in the form of federal and state investment tax credits – a policy that has proven especially beneficial for a very small number of large financial institutions. However, before turning to an examination of the role of tax credits in the solar industry, it is useful to provide an overview of the solar development process.

The solar development process

Solar project development starts with the search for land. An ideal location will have certain physical qualities – clear of brush, not too steep, and away from wetland areas that might spur additional environmental permitting – but equally important is that the land is near a point on the distribution or transmission grid with adequate 'grid capacity', meaning that it can accept additional power without costly line upgrades. In the early years of solar development in the US, grid capacity was less of a concern, as projects were smaller and could therefore be hooked onto the much more extensively available distribution grid. However, as solar projects have grown larger and therefore need to be connected to the transmission grid (i.e. the high-capacity wires that comprise the 'interstate highways' of the electricity grid), ideal locations have become much scarcer and costlier for developers.

Once a developer identifies a suitable property and is able to acquire land – a task that often includes knitting together multiple parcels of land from a variety of landowners – project developers begin designing and engineering the system, acquiring environmental and land use permits and entering into the interconnection queue. While the rules governing interconnection vary across individual states in the US, in general they require the interconnection authority to study whether the local grid can accommodate the extra generation capacity a project might provide and if not, to determine the costs a project developer may be required to pay for transmission upgrades, a sum that can reach well into the millions of dollars. This solar development process is a lengthy one, taking an average of five years from the earliest development work to the point that a project starts pumping kilowatts onto the grid.

For the purposes of this chapter, it is important to note two additional aspects. The first is that few early stage solar developers have any intention of actually owning and/or operating the completed solar farm. Instead, most developers aim to bring a project through portions of the development process – acquiring the land and some permits, for example – and then selling that project to a middle- or later-stage developer better positioned to move the project to the next stage. At the same time, as the project moves through the various stages of development, the risk – and relatedly, reward – of the project changes. An early stage project is still a fairly risky endeavour, as there are a variety of binary risks (i.e. no permit granted equals no project completed) that can lead to cancellation. A later stage project is much less risky.

The second – and related – aspect is that the financial firms that invest at the various stages of development have different goals and risk tolerances. For example, private equity firms are becoming active in early stage development where the risks – but also rewards – are highest. However, by the time the long-term owners of a solar farm decide to invest, the project is nearly operational, meaning that the entire financial lifetime of a solar project can be modelled and predicted with considerable accuracy. In other words, investing in a nearly operational solar farm is a safe – though less lucrative – investment. Yet, as section 4 explains, a solar project produces more than just revenue from the sale of electricity. Rather, because the federal government has sought to incentivise solar projects through the use of tax credits, a project can also generate millions in tax avoidance. However, solar project developers are typically not in a position to benefit from these tax credits. This has led to the formation of tax equity partnerships, a form of electricity capital I turn to next.

Tax equity partnerships

As mentioned, solar projects generate a variety of financial benefits. The first, and most straightforward, is the revenue that comes from the sale of electricity to the offtaker – that is, electricity that is sold on a per kilowatt hour (kWh) basis. Second, in some states renewable energy projects generate renewable energy certificates (RECs) that can be sold to buyers looking to offset their use of fossil fuels (i.e. companies with 100 per cent renewable pledges). Third, and perhaps most important to the growth of renewables in the US, is that solar projects generate substantial tax benefits, in large part stemming from the use of the investment tax credit. The value of the investment tax credit has fluctuated between 20–30 per cent since its introduction in 2005, but with the passage of the IRA the tax credit has been set at 30 per cent of the total cost of the solar project until at least 2033.

The value of the non-revenue financial benefits of solar projects can be substantial – in 2021, the project finance firm Norton Rose Fulbright estimated that the combination of federal tax credits with the generous depreciation rules solar developers can use has allowed for a return of 44–49 cents per dollar of capital cost (Martin, 2021). Yet, because solar developers are mostly small operations with limited tax burdens, they are frequently unable to take advantage of the full tax benefits generated by a utility scale solar project. A key aspect of solar development in the US has therefore been the use of investment partnerships between developers and so-called 'tax equity' investors, typically large financial institutions that owe considerable amounts of taxes (such as Bank of America or J.P. Morgan).

Figure 2.1 shows how these 'tax equity partnerships' are typically structured. In this example, the solar project is owned by a project company, which is in turn jointly owned by the project developer and

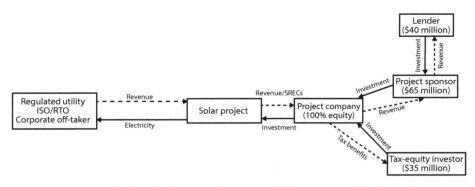

Figure 2.1 Typical tax equity partnership structure. © The author

the tax equity investor. The project developer holds the larger proportion of the project (~65 per cent), while the tax equity investor owns the remainder (~35 per cent). The key aspect of the partnership is that the financial benefits are split between the project sponsor and the tax equity investor. In a 'partnership flip' – a common solar ownership arrangement that is governed by rules from the Internal Revenue Service – 99 per cent of the income, losses, and particularly tax credits accrue to the tax equity investor until they hit a particular yield target. The revenue sharing agreement is structured so that the yield target is hit when all the tax benefits have been captured (usually around five years into the arrangement), at which point the revenue sharing agreement 'flips'. At the point of the flip, the tax equity investor's share of financial benefits drops to 5 per cent, with the project developer taking the remaining benefits (which are now primarily revenue from the sale of electricity and, if applicable, renewable energy credits). It is also typical at that time for the tax equity partner to sell their interest in the solar project.

While commonplace, tax equity partnerships are legally complex and therefore add significant costs to the development of a solar project. In addition, there are a very limited number of tax equity investors, estimated to be as few as 12–15 investors. The result, as Knuth (2021) makes clear, is that tax equity investors can exert outsized control over the types of solar projects being developed. Tax equity investors have a clear preference for larger projects that lower transaction costs and use the relative scarcity of tax equity investment dollars to extract additional fees and outsized returns in exchange for their participation. In short, the US federal government's approach to the renewable energy transition has been, to date, dependent on the continuing appetite and need for tax avoidance among the wealthiest financial organisations in the US.

The IRA of 2022 alters the rules governing the investment tax credit in several ways. Most of these alterations are intended to draw new players into the solar market. These include large investor-owned utilities, most of which had previously only bought renewable energy via power purchase agreements (rather than developing their own projects), as well as public power and other non-profit utilities that paid no taxes and were therefore unable to take advantage of federal tax credits. While the precise ways that the tax credits have been reformed are still being worked out by the Internal Revenue Service as the relevant federal authority, some of the broad implications for solar developers, electric utilities, and investors are becoming clearer as of mid-2023.

The first alteration that comes from the IRA is the introduction of direct pay, a model whereby qualified non-profit entities will receive their tax credits as 'direct pay' rather than something filed with their taxes. The benefits of direct pay are expected to be limited, with only tax exempt entities such as rural electric cooperatives, the Tennessee Valley Authority, local and state governments, and tribal governments able to take advantage (Martin, 2022). The second alteration – called 'transferability' – is a mechanism that allows developers to sell the tax credits to another entity that does not have an ownership stake in the project. In comparison to direct pay, the expectations for how transferability might reshape the solar industry are much broader.

Currently, the tax equity investor is the lynchpin of the capital stack – if there is no tax equity investor, a project will not happen. As a result, the limited number of tax equity investors will only enter partnerships for the largest and safest projects. However, by making solar investment tax credits transferable, tax credits will be sold in marketplaces (many of which are currently being developed) that are accessible to a range of smaller entities that are looking to buy tax credits but not large enough to enter into tax equity partnerships. The expectation is that transferability will also reduce the transaction costs of monetising tax credits (Pizer and Aldy, 2022) and financiers active in the solar sector argue that transferability will help solar developers by opening up new and perhaps more creative methods of financing projects, lowering underwriting standards and broadening the universe of tax investors (Riester, 2022; Rubinstein, 2023).

However, while much of the early focus on transferability has been on the benefits to smaller solar developers, some of the primary beneficiaries are likely to be large investor-owned utilities, many of which have been under pressure from credit ratings agencies to clean up their generation portfolios. Traditionally, tax laws have meant that utilities could glean little benefit from the renewable energy tax credits and thus had limited gain from investment in renewable energy (Varadarajan et al., 2021). However, the transferability of tax credits provides many utilities with the ability to capture multiple financial benefits from solar and not just sell the power at low prices. In the post-IRA world, regulated utilities (which still dominate large portions of the US) can own renewable energy assets, add them to their rate base and therefore make a roughly 10 per cent return on the capital they have invested. At the same time, utilities will also be able to *sell* the tax credits they generate every year, thereby avoiding the situation like the one facing Florida-based utility NextEra Energy that in mid-2023 has $4.3 billion of tax credits it has been unable to use. The result is that a large portion of the interest in tax credit transferability thus far has come from large investor-owned utilities (Good, 2022; Lipton, 2022) and not necessarily from the small developers that seemed most likely to benefit.

Conclusion

This chapter demonstrates the ways in which attention to the cognate aspect of finance can open up new perspectives on how, where and why solar energy transitions are taking place in the US and crucially, whom they benefit. This focus highlights how the quest for and ability to realise profits, and not necessarily the falling price of solar powered electricity generation, has been the key driver of solar development in the US. The recently passed IRA seems set to accelerate the solar industry and deepen the industry's dependence on tax investors to underpin new solar development.

The chapter also demonstrates the benefits of understanding changes in the electricity sector through the framing of electricity capital (Luke and Huber, 2022). Taking a longer view on the finance–state–electricity nexus also draws our focus to notions of 'combined development' (Bridge and Gailing, 2020: p.1041), that is, how remnants of the old energy system carry over and come into conflict and/or combination with new aspects of energy. In this view, the importance of tax policy in the nascent solar industry demonstrates that the US energy transition is not simply about the rational replacement of one least cost or otherwise preferable form of energy generation with another. Rather, the transition has thus far been about how existing modes of investment are finding new opportunities to realise profits.

Finally, the financialisation points to the likely winner in the competing narratives around solar described by Baker (2022). Solar in the US seems much less likely to exist as a decentralised technology that can enhance energy democracy. Rather, even with provisions in the IRA that would purportedly benefit smaller developers, solar in the US has emerged as simply a cleaner and lower carbon tool for private capital accumulation. Larger players remain poised to benefit from their ability to efficiently navigate the at-times labyrinthine tax rules of the IRA. Even at the household scale, a tax credit–driven solar expansion systematically excludes households that do not own their homes – a group that is roughly 34 per cent of the US population and disproportionately composed of people of colour. In short, in the realm of solar we see a variety of investment strategies more organised around the financial benefits of solar – namely tax avoidance – than aligned with any ecological or democratic ones.

References

Baker, Lucy. 2022. Procurement, finance and the energy transition: Between global processes and territorial realities. *Environment and Planning E: Nature and Space* 5(4): 1738–64. https://doi.org/ 10.1177/2514848621991121

Bridge, Gavin, and Gailing, Ludger. 2020. New energy spaces: Towards a geographical political economy of energy transition. *Environment and Planning A: Economy and Space* 52(6): 1037–50. https://doi.org/10.1177/0308518X20939570

Brown, Deward C. 1980. *Electricity for Rural America: The fight for the REA*. Westport, CT: Greenwood Press.

Christophers, Brett. 2015. The limits to financialization. *Dialogues in Human Geography* 5(2): 183–200. https://doi.org/10.1177/2043820615588153

Christophers, Brett. 2022. Taking renewables to market: Prospects for the after-subsidy energy transition. *Antipode* 54(5): 1519–44. https://doi.org/10.1111/anti.12847

Christophers, Brett. 2023. Opinion | The unfortunate, unintended consequence of the Inflation Reduction Act. *The New York Times*, 8 May. Available at: https://www.nytimes.com/2023/05/08/opinion/inflation-reduction-act-global-asset-managers.html (accessed 8 May 2023).

French, Shaun, Leyshon, Andrew, and Wainwright, Thomas. 2011. Financializing space, spacing financialization. *Progress in Human Geography* 35(6): 798–819. https://doi.org/ 10.1177/0309132510396749

Froud, Julie, Johal, Sukhdev, Papazian, Viken, and Williams, Karel. 2004. The temptation of Houston: A case study of financialisation. *Critical Perspectives on Accounting* 15(6): 885–909. https://doi.org/10.1016/j.cpa.2003.05.002

Good, Allison. 2022. Utilities taking up new tax tools in US climate package to finance renewables | S&P Global Market Intelligence. Available at: https://www.spglobal.com/marketintelligence/en/news-insights/latest-news-headlines/utilities-taking-up-new-tax-tools-in-us-climate-package-to-finance-renewables-72373660 (accessed 16 February 2023).

Harrison, Conor. 2013. 'Accomplished by methods which are indefensible': Electric utilities, finance, and the natural barriers to accumulation. *Geoforum* 49: 173–83. https://doi.org/10.1016/j.geoforum.2013.07.002

Harvey, David. 2006. *The Limits to Capital*. London: Verso.

Hausman, William J., and Neufeld, John L. 2002. The market for capital and the origins of state regulation of electric utilities in the United States. *The Journal of Economic History* 62(04): 1050–73. https://doi.org/10.1017/S002205070200164X

Huber, Matt, and Stafford, Fred. 2022. In defense of the Tennessee Valley Authority. Available at: https://jacobin.com/2022/04/new-deal-tennessee-valley-authority-electricity-public-utilities -renewables-green-power (accessed 11 May 2023).

Hughes, Ted. 1983. *Networks of Power: Electrification in Western society, 1880–1930*. Baltimore: Johns Hopkins University Press.

Hyman, Leonard S. 2005. *America's Electric Utilities : Past, present, and future*. 7th ed. Arlington, VA: Public Utilities Reports.

Isser, Steve. 2015. *Electricity Restructuring in the United States: Markets and policy from the 1978 Energy Act to the present*. New York: Cambridge University Press.

Kennedy, Sean F., and Stock, Ryan. 2022. Alternative energy capital of the world? Fix, risk, and solar energy in Los Angeles' urban periphery. *Environment and Planning E: Nature and Space* 5(4): 1831–52. https://doi.org/10.1177/25148486211054334

Knuth, Sarah. 2021. Rentiers of the low-carbon economy? Renewable energy's extractive fiscal geographies. *Environment and Planning A: Economy and Space*: 0308518X2110626. https://doi.org/10.1177/0308518X211062601

Krippner, Greta R. 2005. The financialization of the American economy. *Socio-Economic Review* 3(2): 173–208.

Langley, Paul. 2018. Frontier financialization: Urban infrastructure in the United Kingdom. *Economic Anthropology* 5(2): 172–84. https://doi.org/10.1002/sea2.12115

Lipton, Eric. 2022. With federal aid on the table, utilities shift to embrace climate goals. *The New York Times*, 29 November. Available at: https://www.nytimes.com/2022/11/29/us/politics /electric-utilities-biden-climate-bill.html (accessed 16 February 2023).

Luke, Nikki, and Huber, Matt T. 2022. Introduction: Uneven geographies of electricity capital. *Environment and Planning E: Nature and Space* 5(4): 1699–715. https://doi.org/10.1177/25148486221125229

Mahaffey, James. 2010. *Atomic Awakening: A new look at the history and future of nuclear power*. New York: Pegasus Books.

Martin, Keith. 2021. Partnership flips: Structures and issues | Norton Rose Fulbright. Available at: https://www.projectfinance.law/publications/2021/february/partnership-flips/ (accessed 19 July 2022).

Martin, Keith. 2022. Searching for opportunities in the Inflation Reduction Act | Norton Rose Fulbright. Available at: https://www.projectfinance.law/publications/2022/august/searching -for-opportunities-in-the-inflation-reduction-act/ (accessed 3 February 2023).

Pizer, William A., and Aldy, Joseph E. 2022. Considerations for transferability of tax credits under the Inflation Reduction Act. Available at: https://www.rff.org/publications/testimony-and -public-comments/considerations-for-transferability-of-tax-credits-under-the-inflation -reduction-act/ (accessed 3 February 2023).

Riester, David. 2022. The end of tax equity as you know it. Available at: https://segueinfra.com /articles/the-end-of-tax-equity-as-you-know-it (accessed 6 February 2023).

Rubinstein, Eric. 2023. Clean energy tax credit transferability: Hero of the Inflation Reduction Act? Available at: https://www.utilitydive.com/news/clean-energy-tax-credit-transferability-hero -of-the-inflation-reduction-ac/639690/ (accessed 3 February 2023).

Spinak, Abby. 2020. 'Not quite so freely as air': Electrical statecraft in North America. *Technology and Culture* 61(1): 71–108. https://doi.org/10.1353/tech.2020.0033

Varadarajan, Uday, Posner, David, Mardell, Sam, and Mendell, Russell. 2021. Simple tax changes can unleash clean energy deployment. Available at: https://rmi.org/simple-tax-changes-can -unleash-clean-energy-deployment/ (accessed 6 March 2023).

3

Energy communities as models of social innovation, governance and energy transition: Spanish experiences

Teresa Cuerdo-Vilches

Introduction

Energy communities can be a great support for neighbourhoods, given the current energy crisis situation. Geopolitical conflicts such as the ones in Ukraine and in Palestinian territories and the progressive increase in tariffs due to tax policies on fossil fuels are part of a more complex problem. With an uncertain economic scenario, the many families that already cannot pay their energy bills will have a much more difficult time covering their most basic needs. In a context like Spain, with high annual solar insolation rates, it seems opportune to take advantage of these circumstances to establish collective mechanisms that obtain, manage and store solar energy from spaces, buildings and public infrastructures in the immediate environment. The main aim is to obtain a collection area such that the maximum possible demand can be covered, either immediately or deferred through storage, or by net-metering which reduces the total amount of the electricity bill through prosuming. This chapter reviews the state of the matter at the national level, focusing on solar energy communities as a cognate aspect of solar energy transitions. It examines existing solar energy communities for their most relevant experiences and lessons learned. Through these examples, it is possible to reformulate designs, address technical and operational issues, identify associated risks and impacts, and establish recommendations for public policies that yield social benefits.

Regarding the concept of what an 'energy community' is, there is no specific or unified definition, given the heterogeneous nature already shown in Chapter 1. One of the issues that must be defined is the scale at which it is determined, which can be from a neighbourhood or micro-grid, to a virtual gathering of people. Its members do not have to necessarily be located close to each other, but linked to the same energy co-management purpose. To have an official (as a regulatory reference) definition of what an energy community is, we can find an influential specification in the European Commission's Renewable Energy Directive. This defines up to three meanings for this concept:

a) according to the national law, it is based on open and voluntary participation, autonomous and effectively controlled by shareholders or members located in the proximity of the renewable energy projects, owned and developed by that legal entity; b) shareholders or members of which are natural persons, small and medium enterprises, or local authorities, including municipalities; c) the primary purpose of which is to provide environmental, economic or social community benefits for its shareholders or members or for the local areas where it operates, rather than financial profits. (European Union, 2018)

Another of the most widespread emerging concepts is that of community energy, which refers to a broad set of collective energy actions. These actions relate to the ways in which people may participate in an energy community. As a cognate aspect of solar energy transitions in focus here, solar energy communities generally feature the property that citizens acquire a proactive role with respect to the energy system. The definition of energy communities is a novelty that has been included in European legislation and that the Clean Energy Package recognises as a series of categories of different collective configurations for its production and management, as described later (Caramizaru and Uihlein, 2020).

Many configurations are supported as energy communities, with certain peculiarities. On the one hand, virtual electricity production plants can be found where a group of low-power producers or generating entities come together, grouped as a single market participant. On the other hand, energy hubs are defined by also having an energy management system that allows members of the network an optimised use of energy. Other more complex configurations are integrated community energy systems, which are considered micro-grids, with energy vectors other than electricity, and which are managed by the community itself. Other variations are, for example, the community of prosumers (that is, consumers and producers at the same time), which in addition to producing energy and streaming it to the network, uses it internally efficiently; or the zero-energy community, which is made up of zero-energy buildings that have zero energy consumption at the end of the year. Something similar is the eco-community definition, with the aim of becoming self-sufficient. Some of the benefits of creating this type of organisation are avoiding the growth and associated costs of electric grids while increasing the direct involvement of and benefits to users. Hence, it is possible to reduce reluctance to this social innovation by publicising its benefits.

This chapter reflects upon energy communities as a cognate aspect of solar energy transitions in terms of social innovation and governance. While we can understand cognate aspects as cross-cutting elements that co-shape and co-define solar energy transitions, relevant nuances in this regard vary from one solar energy community to another, due to diverse factors, making the influence that they exert relatively varied. The chapter takes up the Spanish experience, identifying insights from recent praxis. Thereafter, it discusses what such an understanding of energy communities means for solar energy transitions elsewhere.

Energy communities, social innovation and governance

Energy communities are a very important intervention to socially introduce new habits, as well as to change social behaviours. They involve people and make them co-responsible in the task of decarbonising built environments, towards a suitable ecological and energy transition. This has the added value of governance in terms of energy, which also favours other positive side effects, such as autonomy and energy independence, which in times of conflict with the consequent instability of energy supplies, constitute both an opportunity and an urgency to attend to. This independence can enhance energy governance in ways not found in other types of energy management. Non-tangible issues may affect the individual as an active part of the community. The sense of belonging to the collective for a common good, beyond individual benefit, may enable social integration, reduce social inequalities and discrimination, and boost the right to energy. Optimised network management may provide the most disadvantaged members with resources, without necessarily affecting others. More intelligent use of the energy resource, not so much at the grid level, but rather with adequate training and education in this regard, can enable a major behavioural shift that facilitates social interest in solar energy transitions.

The democratisation of energy resources and their management, as well as decision-making, may enhance social cohesion through inclusive, engaging and therefore fair democratic models. Such collective structures also enhance the engagement of citizens in two ways. First, by reinforcing positive social normalisation and support for the energy transition, social participation and widening decision-making capacity on energy issues. Second, by increasing commitment to both social as well as environmental causes.

In the European Union Directive on Renewable Energies, approved in 2019, explicit recognition is made of the citizen's ability to get involved in the production of renewable energy through the constitution of energy communities. Thus, people, local administrations and institutions, and small and medium enterprises, can become legal entities to produce this type of generation and may also receive support from other supranational entities, such as the European Union. The remuneration obtained can be reverted to other services to cover the needs of these same communities (European Union, 2018).

This same directive recognises citizens' right to generate, collect, spend and sell energy. This is very important, because the population is legitimised in this sense, going a step further in the fight for the right to energy (Hesselman, Varo and Laakso, 2019). This is established by the United Nations Sustainable Development Goal 7, which aims to achieve universal access to sufficient, reliable, affordable, and sustainable energy by 2030 (United Nations, n.d.).

With regard to training, internal organisation, management itself and ultimately the governance of energy communities, in Spain various organisations advise in multiple ways on the constitution, expansion and consolidation of these entities. One of the cases is that of cooperative societies, whose main function is to guide real or potential communities in the achievement of their related objectives, as well as to share information about already constituted communities, such as inspiring examples or success stories (Red de Comunidades Energéticas S.Coop, 2020).

There are multiple forms of grouping to trigger an energy community project. Some of these formulae entail a legal entity, which allows administrative management of all the permits and procedures necessary to carry out and put such purposes into effective operation. These groups not only have a legal or administrative role to enable the formal constitution of the community or collective representation, but also play a relevant part in inspiring or facilitating action towards actual or potential members. Other groups, such as associations, neighbourhoods or other collective formations in turn can bring together different actors, such as local administrations (among them town halls), groups of individuals, or other types of consortia between interested parties. In the case of Energy Cities, for example, there are more than a thousand town halls located in some 30 countries, which provide multiple documents, help and information, when it comes to establishing new communities or strengthening existing ones. Among these documents, they have prepared a motivational text on the reasons for establishing an energy community from a sustainable perspective, since they cover its three major approaches: social, economic and environmental (Energy Cities, 2021).

With this background in mind, this chapter considers energy communities as a cognate aspect of solar energy transitions, delving into their peculiarities and differences by focusing on their praxis in Spain. This is one of the most rapidly evolving national contexts where energy communities are emerging with a view to upscaling. The case focus aims to ascertain what learning can be gained from these efforts towards the nature of solar energy transitions themselves.

Spanish energy communities as a response to post-pandemic energetic uncertainties and the climate emergency

The European built-up residential stock in general and the Spanish one in particular, has been under great pressure in recent times. Existing homes – largely built prior to the corresponding regulations on thermal comfort and energy requirements and indoor environmental quality – are mostly energy inefficient and not quite comfortable for users (Cuerdo-Vilches and Navas-Martín, 2022). The overexposure to unsuitable environments for prolonged periods can lead to the generation or aggravation of multiple pathologies or diseases and lack of wellbeing. Organisations such as the World Health Organization have clearly defined the main risk factors in the home, an inadequate domestic environment being the trigger for multiple health conditions (WHO, 2022).

This has been especially observed in the COVID-19 pandemic period. During this time, the house became the epicentre for most people, which made us much more aware of the inadequacy of housing for the usual and extraordinary needs that arose (Cuerdo-Vilches, Oteiza and Navas-Martín, 2020). This was especially relevant for those people in a situation of vulnerability (Cuerdo-Vilches, Navas-Martín and Navas-Martín, 2020). In this period, energy consumption skyrocketed, due to prolonged stays at home, making households even more vulnerable and increasing cases of energy poverty (Cuerdo-Vilches, Navas-Martín and Oteiza, 2021).

Taking into account the current climate emergency scenario, coupled with the energy and geopolitical crisis, it seems more opportune and urgent to displace the current more traditional energy models towards others where fossil energy yields to renewable energy, without losing sight of respect for natural environments and habitats. Models for such ecological and energy transitions are facilitating energy self-sufficiency and independence in environmentally-friendly conditions. Hence, the novelty also lies in the management of these resources itself, in addition to the ability to generate them.

In recent years, many examples of community energy have proliferated in Spain. They have been boosted by the pandemic circumstances and geopolitical events, besides the climate emergency, as well as by European and national initiatives towards building energy efficiency, energy transition and decarbonisation. Hence, the promotion of energy communities has spread nationwide.

Some examples stand out especially for the socio-technical approach they employ. They understand energy communities as essential support for the related aspects that the creation, strengthening and consolidation of the energy community itself entails in the race towards the solar energy transition. The example described below is the RESCHOOL project (http://www.reschoolproject.eu, Horizon Europe grant 101096490), whose social approach makes use of all kinds of strategies and techniques to inform, motivate, share responsibilities with and engage individuals and households as part of energy communities or potential members.

In turn, these strategies, techniques and actions can be grouped according to their purpose. In the first place, some of these actions are aimed at informing. To attract potential members of the energy community, initial training is essential, as well as providing real, adjusted, updated and intelligible information. This allows that people can better understand the energy indicators, in order to feel like an integral part of the system, to be able to make decisions and to see the results of their own actions as prosumers and active consumers. Thus, these households can become relevant actors, capable of proactively participating in electricity markets.

In addition to training for learning as proactive actors, strategies aimed at motivating users are established, based on game elements, such as gamification and serious games. The success of the implementation of these dynamics, mechanics and components of games for non-ludic purposes, relies upon immersing participants in memorable experiences, so that they find a balance between the skills learned and the autonomy achieved, while exercising freedom as users. This feeling of engagement makes people more prone to establishing a bond and commitment. Some of the most common tactics can be the personification of the prosumer or active consumer profile; progress bars of specific knowledge on the subject; competitive or collaborative rankings between users belonging to the network; as well as other drivers or rewards that can be designed to hold the interest of members of the energy community. In turn, serious games can be established for training people safely in the face of possible contingencies or simulated situations and to acquire more knowledge or skills for which they can be rewarded in different ways. These and other techniques are employed to turn users into active prosumers, ready to consume, store or sell to the energy market according to their own needs and those of the sector.

To interact, collaborative community platforms are developed, which complement these more individual or household actions, so that solutions can become more effective, efficient and cohesive with the energy community. On the other hand, this community platform can provide information in real time, which helps decision-making both at the individual, household and community levels, according to the energy input and output flows at different levels.

With regard to technological applications at the service of energy communities, it is worth noting those aimed at optimising and efficient use of energy. These range from modelling focused on making consumption more flexible, to forecasting services for production and demand, through network balancing, the adhesion of new users, or time optimisation services for use and energy management and to the usage of supports and interconnected platforms to plan, visualise and manage all users and bidirectional flows of generation and consumption or dump in the network. All these technological advances also facilitate decision-making based on joint information and predictive models that allow effective use of energy depending on various factors, such as generation capacity, self-consumption, and external demand from the general network or from other actors such as the energy markets themselves.

Another example of the growing Spanish experience in energy communities funded with European projects is the LIGHTNESS project (https://www.lightness-project.eu). In this case, there are participants from five countries, where Spain contributes two very different types of community. On the one hand, there is an energy cooperative created in a town near Valencia, called Alginet. This electrical cooperative was initially created in the first third of the twentieth century to supply electricity to the city due to the disinterest of electrical distributors in electrifying small towns. This cooperative also had an initial social function, since they redistributed the benefits among end users, aimed at mitigating energy poverty, promoting leisure and sports among people, as well as a more recent emphasis on supporting the energy transition. This type of social formation has now found a new way to evolve towards the achievement of these initial objectives, constituting a local distribution system operator, as well as a demand response management system. This allows the development of a strategy to transition to an energy services company, also empowering its end users. The main challenge of being part of this project for this cooperative was twofold: to eliminate barriers in the Spanish energy regulatory framework for small producers and to strengthen the ties between the members of the cooperative and promote proactive consumption and prosumption.

On the other hand, the project has ManzaEnergía as another Spanish participant. This conglomerate was formed from three well-differentiated stakeholders: an energy office in the Manzanares el Real City Hall building (Madrid), the only public school in the municipality (which participated on equal terms with the Manza programme to raise awareness on energy issues) and the residents of the energy community themselves. In 2022, the city council of this Spanish population financed the installation of solar PV on the roofs of the municipal sports centre worth almost 100 kilowatt peak (kWp). This served to generate enough electricity for the sports centre, as well as for the public school and vulnerable households in a situation of energy poverty. A battery was installed to store surplus production. By the end of 2022, residents had formed a civil association to produce and share renewable energy.

In short, more educated, conscious, co-responsible and committed energy communities are capable of making more informed, effective, efficient and, in the end, more accurate decisions. In addition, they perceive themselves as an active part of a collective effort for the common good beyond individual or household co-benefits. This enables not only smooth functioning and improved energy performance towards energy autonomy and independence, but also guarantees a better understanding within and therefore cohesion among community members, which in turn favours better decisions and a more promising and lasting future of this collective.

Energy communities for solar energy transitions

From the two snapshots presented above and many endeavours along similar lines emerging in and beyond southern Europe, it is important to learn lessons applicable to other emerging or lagging contexts of ecological and energy transition. First, the power of people as individual beings and their ability to transform reality and contribute to a paradigm shift must not be underestimated. In moments of uncertainty, the resilience of people must be used to clearly visualise what is necessary, urgent and opportune. Three years after the pandemic, with an energy war waged at the worst possible moment – and a more pressing climate emergency than ever – in addition to other collateral problems such as the scarcity and increase in the cost of basic goods such as water or food, society seems to have nonetheless taken a step forward towards energy independence, the rejection of fossil energy, renewable energy alternatives and the constitution of alternatives for the production, management and distribution of energy. There are many co-benefits linked to these forms of constitution of self-managed collectives. Some derive directly from the feeling of belonging and are fuelled by the motivation to feel proactive and useful, and to perceive results not only at an environmental level, but also embodied in one's own energy consumption and reduced energy costs.

In this sense, community organisation entails a greater production capacity, a higher impact on participating households, and a greater capacity for inclusion than current energy configurations, reducing social inequalities and vulnerability to energy poverty. This also makes possible adaptation strategies to the climate emergency and change, by giving families greater choice and moving them a little further away from poverty.

However, whether administrative, institutional, governmental and regulatory modes of organisation can accomplish fluid interaction can be decisive for the success of energy communities, as it is vital for avoiding gaps or barriers to their creation. In Spain, authority over energy decisions rests largely with the autonomous communities and is thus exercised in a decentralised manner, with unequal treatment throughout the country. This complicates bureaucratic, regulatory and managerial matters both for promoting solar energy and to incentivise and support institutional recognition of emergent forms of energy associations in civil society. In this respect, the role of the central government combines with socioeconomic and cultural factors to influence people's perceptions and attitudes. Residents of regions where the most basic needs such as education or employment have not yet been adequately met find it difficult to worry about environmental aspects. It is thus of vital importance to promote the multiple advantages of constituting energy communities, not only as a way of mitigating vulnerability and democratising energy, but also to create jobs and reactivate an economy that has been through several major crises in recent years.

In turn, the social aspect is closely linked to these starting conditions. The population must be aware of these collective solutions to produce, share, consume or sell energy and the multiple advantages derived from this alternative to conventional energy distribution. For this, a whole network of well-organised stakeholders is necessary, aimed at supporting, informing, incentivising, collaborating within and even investing in these types of initiatives. These networks may be formed by public administrations, political decision-makers, legal instruments, as well as municipalities and the leaders of potentially participating public entities (for example, health, sports, educational centres or many others well-positioned to cede their roofs and surfaces for energy production). Moreover, other stakeholders such as industries, experts, researchers and academics, and the general public may join energy communities. All these interested parties must have accessible, simple, up-to-date and customised tools that facilitate access to functional energy community frameworks, with clarity over the rights, duties, warranties and responsibilities of each party. Otherwise, information confusion, legal loopholes, administrative complexity and other barriers can lead to demotivation, discredit and, finally, abandonment.

Lastly, specific training in energy, generation, distribution, balance, energy markets and related issues is of vital importance for end users and potential prosumers. People have to find clear co-benefits that must be highly promoted and to understand issues such as vulnerability mitigation, alongside learning to manage their own consumption and demands and taking on the role of highly relevant decision-makers. Thus, education, critical awareness, a sense of co-ownership and engagement are essential for the paradigm of the energy transition and decarbonisation to take hold. All these reasons make energy communities a relevant and definitive cognate aspect of solar energy transitions.

References

Caramizaru, Aura, and Uihlein, Andreas. 2020. *Energy communities: An overview of energy and social innovation* 30083. Luxembourg: Publications Office of the European Union.

Cuerdo-Vilches, Teresa, and Navas-Martín, Miguel A. 2022. Geo-Caracterización Energética de La Vivienda Cordobesa: Aplicación de Clústeres Aproximativos a Escala Municipal. *WPS Review International on Sustainable Housing and Urban Renewal* 11–12: 111–28. https://doi.org/10.24310/wps.vi11-12.15906

Cuerdo-Vilches, Teresa, Navas-Martín, Marina, and Navas-Martín, Miguel A. 2020. Estudio [COVID-HAB-PAC]: un enfoque cualitativo sobre el confinamiento social (COVID-19), vivienda y habitabilidad en pacientes crónicos y su entorno. *Paraninfo Digital*: e32075o–e32075o. Available at: https://ciberindex.com/index.php/pd/article/view/e32075o

Cuerdo-Vilches, Teresa, Navas-Martín, Miguel A., and Oteiza, Ignacio. 2021. Behavior patterns, energy consumption and comfort during COVID-19 lockdown related to home features, socioeconomic factors and energy poverty in Madrid. *Sustainability* 13(11): 5949. https://doi .org/10.3390/su13115949

Cuerdo-Vilches, Teresa, Oteiza, Ignacio, and Navas-Martín, Miguel A. 2020. Proyecto sobre confinamiento social (covid-19), vivienda y habitabilidad [COVID-HAB]. *Paraninfo Digital*: e320660–e320660. Available at: https://ciberindex.com/index.php/pd/article/view/e320660

Energy Cities. 2021. Comunidades Energéticas: Una Guía Práctica Para Impulsar La Energía Comunitaria. Energy-Cities.Eu. Available at: https://energy-cities.eu/wp-content/uploads /2020/10/guia-comunidades-energeticas.pdf

European Union. 2018. Directive (EU) 2018/2001 of the European Parliament and of the Council of 11 December 2018 on the promotion of the use of energy from renewable sources. Available at: http://data.europa.eu/eli/dir/2018/2001/oj

Hesselman, Marlies, Varo, Anais, and Laakso, Senja. 2019. European energy poverty: The right to energy in the European Union. Available at: http://www.engager-energy.net/

López-Bueno, J., Díaz, J, Sánchez-Guevara, C., et al. 2020. The impact of heat waves on daily mortality in districts in Madrid: The effect of sociodemographic factors. *Environmental Research* 190: 109993. https://doi.org/10.1016/j.envres.2020.109993

Red de Comunidades Energéticas S.Coop. 2020. Red de Comunidades Energéticas S.Coop. - Comunidades Energéticas. 12 June. Available at: https://comunidadesenergeticas.org/

United Nations. n.d. Ensure access to affordable, reliable, sustainable and modern energy for all. Available at: https://sdgs.un.org/goals/goal7

WHO. 2022. *Directrices de La OMS Sobre Vivienda y Salud*. Pan American Health Organization. Available at: https://iris.paho.org/handle/10665.2/56080

4

Beyond power: the social situatedness of community solar energy systems

Karla G. Cedano-Villavicencio
and Ana G. Rincón-Rubio

Introduction

The concept of 'situated knowledge' emerges from subaltern racial-ethnic and feminist epistemological perspectives, exemplified by scholars such as Donna Haraway, Gloria Anzaldúa, Adriana Guzmán, Enrique Dussel, Walter Mignolo and Ramón Grosfoguel. They have criticised hegemonic positivist and Eurocentric paradigms that claim to possess a universal, neutral and objective standpoint. These critical epistemologies assert that all knowledge is inherently 'situated'. In other words, knowledge production is always rooted in a specific location within power structures, placing it within the dominant or subaltern side of power relations. No one is exempt from the influence of racial, class, sex, gender, spiritual, linguistic and geographical hierarchies (Grosfoguel, 2006). Consequently, the notion of disembodied and delocalised neutrality and objectivity in knowledge is a fallacy (Grosfoguel, 2006).

Decolonial and feminist thinkers, including those mentioned earlier, emphasise that acknowledging our standpoint not only does not compromise the rigour of research but, conversely, offers valuable insights for a more profound understanding of social realities. This acknowledgment is essential for a positive transformation of society. Therefore, our explicit intention is to position ourselves openly – 'situarnos' – to contribute to constructing critical and rigorous knowledge about solar energy. As young Mexican academic women, our contribution has a declared aim of benefitting those who the modern colonial civilisation system has marginalised and oppressed.

In this chapter, we present two case studies from Mexico to illustrate a new take on solar energy projects. Our perspective frames the cognate aspect of social situatedness. Based on these two case studies, the objective is to identify some hidden agency layers of energy interventions that may be common to those implemented in other contexts and whose understanding must be deepened to achieve energy transitions with a sense of community.

Learning from past experiences where a science, technology, engineering, and mathematics (STEM)-focused approach to the design and implementation of solar energy solutions has all too often dominated, these two case studies reveal the importance of taking a wider view on solar energy. The first showcases a middle-out approach, where early career social sciences and humanities (SSH) and STEM scholars promoted solar photovoltaics for electricity (PV systems), heating and dehydration. Three different processes that use solar energy, to three different peri-urban, energy-poor, women-led households, targeted at their specific productive needs. The second one entails the analysis of two community solar PV projects that were installed in rural, isolated, energy-deprived indigenous communities; the decay of the technology and the feelings and emotions these transitions unveiled. In both cases, people's social and energy reality contrasts with the government narratives about the population's actual connection to the power grid and the use of fuels for thermal applications.

We briefly describe the two case studies and build from the interdisciplinary view on their results, impacts and lessons learnt. We highlight the need to address solar energy systems as social, sustainable and inclusive systems, seizing the diversity that solar technology offers to alleviate diverse energy needs. Analysing power dynamics and emotions at the community and household level and given the extent to which solar technology (thermal and PV) is seen as a vehicle to improve quality of life, we mainframe and unpack the relationship between impacts and the degree to which solar projects are conceived of in tune with the needs and affinities of people and their communities. This draws out the significance of social situatedness as a cognate aspect of solar energy transitions at the community scale.

Renewable energy research organisations as a motor of change

Every day in Mexico, more and more people are interested in moving towards a new energy system model – based on renewable energy sources – that is more efficient, accessible, decentralised and one that also contributes to the security and democratisation of energy. Although the macrostructural political and economic conditions prevent a radical change of the energy model in the short term, the groups that support an energy and social transformation, many of them emanating from academic spaces, have adequate agency to promote decentralised projects that can be replicated in various localities of the country, especially those that suffer from greater social exclusion.

An example is the Community Empowerment through Solar Energy project promoted by the most recognised and largest public university in Mexico, which was developed in Temixco, Morelos, in the central region of the country. This project was promoted in peri-urban localities, which, although they already had a connection to the power grid, did not have access to solar energy to power their productive activities.

A second case is the creation of the Network of Solar Communities, an autonomous solar energy generation project promoted by professors and students from a public university in Chiapas, a border state in southern Mexico, which has the highest national rates for economic poverty (INEGI, 2020) and energy poverty (García-Ochoa and Graizbord, 2016). This project was promoted in rural localities inhabited by people of different ethnic affiliations who did not have a connection to the power grid.

Both experiences were developed from STEM academic spaces that sought to contribute to the development and wellbeing of locations underserved by the state and the market. From these interventions, we were able to identify layers of social agency in energy interventions at the community level, namely a process-focused layer and the emotional layer.

Beyond STEM: the hidden layers of the energy transitions

Process-focused layer

The Instituto de Energías Renovables of the Universidad Nacional Autónoma de México, was founded in 1985 as the Laboratorio de Energía Solar, with a community of STEM academics who carried out basic and applied research activities on PV and thermal energy. As the years passed, the research topics were expanded to other areas of renewable energy and technological development was incorporated into the community's work. Thus, the laboratory evolved to establish the Centro de Investigación en Energía in 1996, only 11 years later. In these years, few advocacy projects were carried out. Even these few were designed and implemented exclusively by STEM scientists with a top-down approach. In 2013, the community achieved an unprecedented evolution in the scientific history of Mexico, by establishing itself as a networked academic entity to address new research topics, the Instituto de Energías Renovables; to develop new forms of multidisciplinary collaboration; to focus its work on innovation; and to incorporate a gender perspective in its academic work. The transition from a vertical and hierarchical entity to one organised in a network, allowed the inclusion of scholars with diverse academic profiles. In other words, although the institute continues to have STEM academics, several people have been trained through experience in administrative, economic and social sciences. In 2017, the research group on Social Demand of Energy was integrated, whose main characteristic is to recognise that energy is not an end, but a means to achieve wellbeing within the framework of sustainable development.

This awareness of the importance of energy systems allowed this group to collaborate with academic entities in the United Kingdom that fostered a different vision to the top-down approaches so popular in technological development projects. In one of our meetings, while looking for future collaborations, we wondered about the role that solar energy plays in empowering people to improve their quality of life in Mexican peri-urban regions. We were especially interested in peri-urban regions for two reasons. First, very little work has been done to better understand peri-urban dynamics in Mexico, so that posed an interesting project by itself. Second, most of the public universities are located in peri-urban regions, so we could design a methodology that could be led by local universities with their neighbouring communities.

Peri-urban areas are considered to be on the electric grid. However, the economic reality of the people who live in these regions creates a niche for informal grid access. This informal access is painfully familiar to the Mexican public utilities and is sheltered under the concept of 'non-technical losses' during distribution, which accounts for 6 per cent of the energy generation. Informal access applies also to thermal energy, where the use of liquefied petroleum gas or natural gas is replaced by informal wood management. Energy poverty in these regions poses different challenges in terms of assessing it, understanding it and most importantly, alleviating it.

To address these issues, our main goal was to look at peri-urban locations and characterise their built form morphology and to estimate the potential to use and manage solar energy. By considering a variety of socio-economic conditions and urban settings, we believed it would be possible to characterise the social potential of using and managing solar energy and how it could be used to empower improvements in quality of life and reduce harms along health, environmental, social and economic lines.

A crucial part of the work was to also identify areas of 'change', both in terms of economic development and personal empowerment. Finally, for the project to be replicable, we defined that the methodology that we would develop should be middle-out. That is, it would focus on an academic entity to be the manager of change towards the community and the other stakeholders (local and regional government, non-governmental organisations, suppliers and so on). To pilot this, we worked with three case studies in Temixco, the peri-urban region where the Instituto de Energías Renovables is located. These case studies were selected to foster productive projects in households headed by women. This responded to the nature of peri-urban regions, where informal micro-businesses are located within homes, and where women are the de facto managers of the household and the informal business, in line with the gendered power dynamic that prevails. Thus, three informal businesses were selected: a small typical restaurant (or *fonda*, in Spanish), a tamale shop, and an artisanal cosmetics manufacturer.

Convinced of the need for interdisciplinary work, several young scholars from different fields of knowledge were included in the project design. Thus, 12 young scholars received scholarships, six from STEM areas and six from SSH areas. Within this group, we had a postdoctoral fellow with expertise in educational research and another postdoctoral fellow with expertise in energy engineering. Both shared the operational responsibility for the project along with the lead technical manager. Both teams (the STEM and the SSH) had an equal composition in terms of gender. As a matter of fact, this became one of the strengths of the project when working with female heads of households.

The incorporation of the humanities, represented by the postdoctoral fellow in educational research, enabled us to move away from vertical teaching techniques to sensitise families. Guided by her expertise, we incorporated a process of energy literacy, co-designed by STEM and SSH participants at the initial stage of the project. Thus, despite the COVID-19 crisis, it was possible to work with families and generate an action-oriented understanding of energy and its management. In the same way, the heads of families selected, together with the STEM teams, the solar technology that best met their needs. Thus, the *fonda* owner requested solar water heating, given the need to sanitise the utensils, dishes and cutlery. The tamale cook requested a PV installation that would allow her to reduce the electrical consumption of refrigerators and lighting, to protect her value proposition – freshly made tamales every day. Finally, the cosmetics producer requested a semi-industrial solar dehydrator to prepare the raw material and diversify her product offer.

It is important to highlight that, since the objective was to empower them through the strengthening of their productive projects, the household leads requested a workshop on entrepreneurship. This workshop was delivered by an expert in sustainable and community entrepreneurship, which provided the entrepreneurs with basic accounting tools and more importantly, with a clear understanding of what their value proposition was. In this way, they were able to integrate renewable energy sources in their business models without losing sight of the quality of their products and were fully aware of the economic advantage that a more sustainable business generates in the medium and long run.

This project promoted the possibility of knowing, using and taking advantage of different solar energy technologies, since it included PV and thermal uses to meet productive needs. Since solar thermal is more efficient than PV, understanding the process that required energy was fundamental in the technology selection. The energy literacy process was, without a doubt, the key element of the project. The actionable knowledge that the families acquired enabled them to make energy decisions in the face of technological solutions that they did not know before being part of this project. This understanding increased the energy and economic agency of the people involved, beyond the installation of the technologies.

This would not have been possible without the continuous and constant interaction of the critical and proactive young scholars who, with great alacrity, improved the original design of the project. This helped to transform a pseudo middle-out approach – which was in practice a top-down approach from the academy to the community – into a truly interactive, vibrant and adaptive middle-out approach. Thus, the empowerment project achieved its goal, not only empowering the heads of three families in Temixco, but also creating an empowering space for the entire team, beyond credentials, levels and academic experience.

Emotional layer

In official reports, Mexico boasts that 99 per cent of homes have access to electricity (INEGI, 2018). However, the fieldwork carried out by STEM and SSH research teams found that there are impoverished rural populations that are not even recognised in maps and that do not have a connection to the power grid. Thus, there is a triple invisibility of these social groups: economic, energy-related and territorial. That is, there is a parallel between off-grid and 'off-map'. This second section addresses such a case.

In 2013, people from seven rural and indigenous towns that did not have electricity in Chiapas, a border state in southern Mexico, requested support from researchers from the Instituto de Energías Renovables de la Universidad Pública de Chiapas. Some researchers, characterised by their social commitment, listened to the needs of the communities and designed an off-grid energy project based on solar energy. The project was so innovative and important in terms of technology and social impact that it was recognised in local and state media as constituting a watershed moment in terms of access to energy for isolated populations.

From 2019 to 2021, another small group of researchers carried out a social analysis of this case. In this study in which we partook, we found that the off-grid generation project had an operational lifespan of five years, a period in which people had access to some low-energy consumption appliances. By 2019, the project had stopped working almost completely due to the end of the useful life of batteries and cables, on top of theft and vandalism related to the solar PV panels. Thus, we identified three crucial moments related to energy in the individual and collective life of the people of these regions: a) having lived all or a large part of their lives without electricity; b) having had electricity produced by autonomous generation systems; and c) having to live again without electricity, but now with the previous experience of having had access to it for a few years.

When analysing this process, we found that the arrival of electricity had an extremely positive impact on people's emotions, both individually and collectively. However, when access to electricity was cut off again, emotions such as sadness, disappointment, isolation, anger and guilt surfaced. People experienced these negative feelings more intensely than before they had the experience of living with electricity, since the experience generated a great contrast.

However, this story took a new turn, as anger and deep sadness motivated people to start a new fight for electricity, until they succeeded at the end of 2021. These findings show the ability of affective ties to make possible social processes around energy, allowing us to identify the collective agency and recognising the inhabitants as active stakeholders who can transform their reality. This insight is crucial to understanding and appreciating how the link between emotions and agency works in social processes around energy.

From this analysis, we established the concept of emotional energy communities, which we understand as groups of people linked by shared emotions that have been politicised and have become detonators or catalysts for energy transformations at the community level. The social actors that make up the emotional energy communities are not only the residents who live in the same geographic and energy setting, but also those external agents that are linked to these localities by ties of empathy and collaboration (Rincón-Rubio and Cedano-Villavicencio, 2023).

What has been learnt from this experience is that the emotional impact of off-grid power generation is not unique, stable or unidirectional. On the other hand, this type of project generates mixed emotions and can be contradictory, because access to electricity does not completely solve the energy problem. In sum, this layer refers to the fact that shared emotions around off-grid projects with solar energy can generate new social processes regarding energy management. Shared emotions, whether 'positive' (such as joy and pride) or 'negative' (such as anger and fear), promote bonds of empathy and reinforce a sense of community, thereby enabling collective agency around energy (Rincón-Rubio and Cedano-Villavicencio, 2023).

Lessons learnt: keys to addressing the energy transition with communities

This chapter addressed the situated nature of solar energy transitions by describing two local social processes directly related to solar energy. Based on these cases, some key issues were identified for the analysis of cognate aspects of solar energy transitions. In the first place, we hold that research on energy with practical social implications is very valuable in countries with great natural, territorial, social and economic differences, such as Mexico, because quantitative reports at the national level tend to diminish the realities of these small and medium populations. Showing these experiences from academia is key towards identifying people's agency by counteracting governmental diminution of those who directly interact with energy systems, beyond the big players in the driver's seat, such as the state and the market. Likewise, in the context of the Global South, it is still very important to not only analyse the processes around electricity, but also thermal energy, since it is directly linked to resources such as water, food and the productive activities of rural and peri-urban families.

Furthermore, advocacy projects require the collaboration of STEM and SSH academics and young scholars. This collaboration must be respectful, inclusive and supportive. The outcome of these interventions depends mainly on effective and purposeful communication between all the disciplines, and on active listening, especially by the established academy to trained intersectionality scholars. Energy systems must be conceived as a means for the development of individuals and their communities. For this, social innovation must be redefined as participatory knowledge management to design projects, programmes and public policies that improve the quality of life of individuals and their communities by enabling their freedom of choice (Martínez-Tagüeña et al., 2016).

The experiences described here on processual empowerment and emotional empowerment have allowed us to distinguish two layers of social agency that influence and are influenced by energy systems. It is important to address them to better design energy projects at the community level, mindful of the cognate aspect of social situatedness. Here, we emphasise the process and emotional layers of situatedness.

The collective agency around energy created as a result of the implementation of small and medium solar energy projects, such as those mentioned here, allows us to think that the 'just energy transition' might not emerge from a large state or business intervention. On the contrary, our experience suggests that the energy transition might be the compound interaction of multiple community energy transitions, from a large set of projects designed for and with the communities, respecting the diversity of contexts, habits, beliefs and individualities. This offers a basis to move forward from an abstract, ideal, unique notion of an energy transition to a more realistic, attainable concert of interlaced, diverse, socially situated energy transitions.

Acknowledgements

The support of the Federal Mexican Government's Consejo Nacional de Humanidades, Ciencias y Tecnología (Mexico) is gratefully acknowledged with funding for the project 'CeMIESol P70-Validación de estrategia para Empoderamiento mediante aprovechamiento energia solar (207450)', as well as through the Program for Teacher Professional Development funding project, 'Aportes desde el feminismo y los estudios sobre el territorio a los proyectos energéticos: El caso de la red de comunidades solares en Cintalapa, Chiapas (511-6/2020-8622)'. También, queremos agradecer, respetando su anonimato, a las personas participantes y beneficiarias por su generosidad y confianza durante la investigación de la que fueron parte, sin ustedes este trabajo no sería posible.

References

García-Ochoa, Rigoberto, and Graizbord, Boris. 2016. Caracterización espacial de la pobreza energética en México: Un análisis a escala subnacional. *Economía, Sociedad y Territorio*, XVI (51): 289–337.

Grosfoguel, Ramón. 2006. La descolonización de la economía política y los estudios postcoloniales: Transmodernidad, pensamiento fronterizo y colonialidad global. *Tabula Rasa* 4: 17–46.

INEGI. 2018. Encuesta Nacional sobre Consumo de Energéticos en Viviendas Particulares (ENCEVI). Available at: https://www.inegi.org.mx/programas/encevi/2018/

INEGI. 2020. Encuesta Nacional de Ingresos y Gastos de los Hogares. Available at: http://www .inegi.org.mx/contenidos/programas/enigh/nc/2020/doc/enigh2020_ns_presentacion _resultados.pdf

Martínez-Tagüeña, Natalia, Cedano-Villlavicencio, Karla, and Martínez, Manuel. 2016. Una propuesta de Niveles de Maduración Tecnológica para Ciencias Sociales, V Congreso Nacional de Ciencias Sociales, Chiapas. Available at: https://www.comecso.com/wp-content/uploads /2016/02/Programa-final-5-Congreso.pdf

Rincón-Rubio, Ana G., and Cedano-Villavicencio, Karla. 2023. Emotional energy communities: Centering emotions and feelings within energy transitions in southern Mexico. *Energy Research & Social Science* 98. https://doi.org/10.1016/j.erss.2023.103014

5

Can solar energy make up for a failing grid? Solar energy deployment in urban and urbanising localities of the Global South

Bérénice Girard, Alix Chaplain and Mélanie Rateau

Introduction

Urbanisation is rapidly increasing in the Global South, which is now home to 75 per cent of the world's urban population. By 2025, an estimated 3.75 billion people will live in cities of the Global South (Smit, 2021). While environmental challenges associated with rapid urbanisation, such as water and wastewater management (Björkman, 2015; Bakker, 2010), solid waste (de Bercegol et al., 2017; Cavé, 2015) and air pollution have been extensively discussed in academic literature, the issue of energy usage and practices has received relatively less attention. In the absence of a reliable or affordable grid, urban dwellers in the Global South often use a variety of energy solutions, including informal grid connections, diesel generators, mini-grids, batteries and solar-powered solutions to access electricity (Rateau and Jaglin, 2020). These solutions are frequently combined to ensure a stable energy supply and meet different energy needs. For example, a connection to a local diesel mini-grid can provide backup during an electricity outage, while a solar panel can power a fan or two during the day. Institutions also invest heavily in alternative solutions to ensure 24/7 supply, reduce their electricity bills, or minimise their dependence on the grid (Pilo, 2022). We argue that solar energy transitions are linked to this diversity of alternative energy solutions within the heterogeneous configurations through which urban dwellers in southern countries access energy. Drawing on Lawhon et al. (2017) and Rateau and Jaglin (2020), we define these electricity configurations as dynamic mixes of multiple technologies, uses and systems of actors, which can shift depending on the context, the availability of funds or technical objects, etc. Solar energy is often one of several options in these configurations. It is not always the most affordable or the most adapted to consumers' needs. It is often in competition with or complementary to socially and politically well-embedded alternatives.

Based on extensive qualitative fieldwork conducted in urban and urbanising areas of Northern India, Lebanon and Benin, this chapter situates solar energy in the market for these alternative energy access solutions. In other words, why and how do people living in these localities use solar energy? How does solar energy fit into urban markets for alternatives to the grid? By taking a bottom-up approach to energy practices and use, we aim to inform the complex relationship between the grid, solar energy solutions and other complementary devices in urban areas of the Global South. As a cognate aspect of solar energy transitions, we argue that the heterogeneous configurations through which urban dwellers access electricity in the Global South merit attention. Attending to these configurations can help us rethink the transition to solar energy, beyond large-scale numbers of increasing installed capacity at the national and international scales, to take into consideration people's needs, uses and aspirations in different contexts. A focus on actual practices of accessing energy can also help us identify potential pathways for a more equitable, just, and rapid transition to sustainable solar energy in the urban Global South.

The next section provides the context of study, by examining the challenges to universal access to stable electricity supply in the Global South. The subsequent section describes the different facets of the use of solar energy in the localities and countries of study. In the final section, we discuss what these cases reveal about the future of the energy transition in the Global South.

Failing electricity grids

Large sections of the urban population of the Global South do not have access to an affordable and reliable supply of electricity. Rapid urbanisation and population growth, increasing energy demand, ageing infrastructures or unaffordable connection costs and tariffs are among the main challenges to ensuring universal access. Connecting to the grid can be particularly difficult for the poorest urban residents, such as people living in unplanned settlements and slums, or street vendors. Even when they do have access to the grid, many urban dwellers must face regular electricity cuts and load shedding. The deployment of the grid and the quality of its supply are often highly dependent on economic or political contexts at different scales. Thus, international events, such as the war in Ukraine or attacks on pipelines, can lead to a rapid deterioration of electricity supply. So can national political and economic crises. Structural adjustment policies have also often led to the restructuring and privatisation of electricity utilities and rendered public investments in large electricity infrastructures 'unviable' (Cross, 2013), promoting decentralised, privately-run solutions instead. Regional politics or urban planning policies can also be a factor, leading to wide disparities in electrification rates within a country or a locality. The three contexts of study, which we detail further below, illustrate these different dynamics and the 'uneven temporalities of conventional grids' (Verdeil and Jaglin 2023, 3), such as the rapid development of the grid and improvement of its supply in a specific region (Bihar), the improvement of supply coupled with high connection costs in specific urban areas (Cotonou), or the structural deterioration of grid supply throughout a country (Lebanon).

The state of Bihar, situated in northern India, has long been known for the failure of its basic services and networked infrastructures. In 2011, it had a household electrification rate of 16.4 per cent, compared to 67.2 per cent nationally. This was compounded by unreliable supply, which often did not exceed a few hours per day. Some villages were also considered 'de-electrified' – the electrical infrastructure having ceased to function or having been deliberately damaged (Siddiqui, 2018). In recent years, however, grid access and electricity supply have made great strides, due both to political impetus at the state level and to a general push from the national government to reach universal access, including free electricity connections for poor households. Thus, Bihar has now, at least according to official numbers, reached 100 per cent household electrification, though consumers must still deal with regular load shedding and electricity cuts. In the three urban and urbanising localities of study, residents usually get between 16 and 22 hours of grid electricity a day.

In Benin, a small West African country, cities have an average electrification rate of over 67 per cent, compared to just over 18 per cent in rural areas. More specifically, the authorities estimate that the rate of access to the conventional grid in Cotonou, Benin's economic capital, is 99 per cent. However, this statistic does not reflect the reality of household electrification, as the method of calculation focuses on the locality level.

Electrification programmes are deployed based on a dual reading of land. Urban planning documents contrast planned urbanisation areas with unplanned areas. Unplanned areas are not prioritised in conventional grid extension projects. Therefore, in a context of constant improvement in grid supply, the main challenge that households face is that of connecting to the grid. City dwellers can apply for individual connections at an exorbitant cost. Since they cannot afford it individually, households cobble together informal extensions and work out arrangements with their neighbours, either to buy electricity from a neighbouring retailer or to connect their home to an existing electric meter.

In contrast, in Lebanon, 99 per cent of the national territory is electrified, due to a regional and spatial development policy carried out in the 1960s by then-president Fouad Chehab. The challenge facing residents is therefore not the grid connection, but rather the reliability of supply. Indeed, since the civil war, which ended in 1990, Lebanese consumers have suffered from power cuts ranging from three hours in the capital Beirut to more than 12 hours in remote areas. The situation has further deteriorated since 2021, due to the dramatic economic, political and social crisis nationally. Drastic load shedding has been implemented for all inhabitants, with 22 hours of electricity cuts a day. Although electricity infrastructures were destroyed during the civil war and then again during the 2006 war with Israel, the poor condition of the electricity sector is mainly due to political reasons, including legal controversies, failure to pass a budget and political blockages.

Faced with an erratic supply from the grid, households and institutions in the various localities of study use a diversity of solutions, both individual and collective, to access electricity or have back-up. Among the cheapest solutions are small solar and/or battery-powered lanterns, informal connections to the grid (as in Cotonou) and connecting to a local diesel or solar mini-grid. Better-off households and institutions customarily invest in batteries and diesel generators. Due to technological improvement and cost declines, some also invest in larger on-grid or off-grid solar rooftops. Even when grid supply improves, households and institutions often have low trust in the conventional grid. They thus often keep alternative solutions until they stop working (Cross and Murray, 2018). In this chapter, we consider this sociotechnical heterogeneity to exist 'not as a result of the incompletion of the modern infrastructural ideal of the networked city, but rather as a modality of development in contexts that are marked by social and political fragmentation, great poverty, and new forms of urbanity' (Verdeil and Jaglin 2023, 2). Alternative solutions to the grid have historically been conceived of as short-term, waiting-for-the-grid solutions for poor households and are often ignored by policymakers. In this chapter, we argue to the contrary that to understand current solar energy deployment in urban areas of the Global South, as well as to promote a more rapid, just, and sustainable deployment, one must take into consideration this sociotechnical plurality and the actual energy access practices of households and institutions.

Solar energy as one option among many

In the following paragraphs, all based on extensive qualitative fieldwork conducted by the authors,[1] we illustrate the diversity of solar energy solutions available, the varied drivers for their adoption, diverse usage and emerging challenges in terms of social justice and sustainability in urban and urbanising localities of Northern India, Lebanon, and Benin. We show how the deployment of solar energy is dependent on the presence of the grid, the quality of its supply, the general trust of residents and institutions in networked infrastructures, the evolution of the tariffs of competing energy sources and local technical expertise. All of these modulate the configurations through which households and institutions access electricity wherein solar constitutes one of several sources.

In Bihar, the market for small solar solutions, such as solar home systems and solar lanterns, was booming until recently in a context where the grid was either absent or failing. The great strides made in recent years in improving grid access and supply has however had a deleterious impact on this market. Shopkeepers indicated a rapid decline in sales and underlined that households who could afford it now tended to reinvest in invertor and battery kits, which were cheaper and often better suited to their needs, as they provide a few hours of back-up whenever needed. Households who owned solar solutions would routinely keep using them, but often declared that they were not planning to reinvest in such products if they stopped working.

By contrast, in Benin, the market for distributed solar solutions remains dynamic and the range of solar products available in the markets of Cotonou (Rateau and Choplin, 2022) is sufficiently varied to meet the diversity of needs and purchasing capacities. In areas not connected to the conventional grid, even informal extensions remain beyond reach to some households, due to unsuccessful negotiations with neighbours, costs and delays, or geographical remoteness. Some of these households then rely on solar home systems to access electricity, limiting their uses to a few light bulbs and a fan, more rarely to a television and a refrigerator. These households consider solar energy as an intermediate and temporary solution while waiting for the grid, on which they place their hopes for unrestricted electricity consumption. The households we met adjusted their choice of equipment to the quality/price ratio they could afford. They bought their solar kits from informal vendors or local hardware stores. Generally, solar products are not subsidised, certified or guaranteed, as households that depend on solar devices are excluded both from conventional urban grid electrification policies and from subsidised rural solar electrification policies.

In areas of Cotonou connected to the grid, wealthier households often own diesel generators to provide back-up during the rare power outages. As grid supply improves, the maintenance of these generators becomes less regular. Households prefer to turn to prepaid meters or solar panels. Their objective is not to secure their electricity supply, but rather to control and reduce their energy bills, in a context where they often use several air conditioners. They thus benefit from access to electricity within the frameworks of formality, both for grid supply and solar energy. These wealthier households buy their solar equipment in specialised stores and benefit from the advice of technicians to choose equipment that is adapted to their needs. The products they buy are covered by warranty and sometimes subsidised.

Similar dynamics are emerging in Bihar, where some national brands have started advertising solar home systems not only as back-up solutions, but also as a financially smart investments that could help reduce electricity bills. However, shopkeepers selling these solutions complained about a lack of interest from their clients. Artisans and small industries commonly declared that in the absence of financial support from the state, they did not have the capacity to invest in solar solutions and thus often relied on diesel generators. On-grid solar rooftops started to emerge by the 2020s, but their deployment faced multiple difficulties, including a lack of capacity and expertise at the local level from the public distribution company's engineers as well as from entrepreneurs. In the absence of supportive subsidies and financing solutions, the investment was also beyond most households' financial capabilities and net meters for electricity were often very difficult to get installed. Distribution companies were indeed often reluctant to provide them, as they saw on-grid solar as a potential threat to their revenues.

In Lebanon, state actors in close partnership with international donors have been supporting the development of solar energy through market-based mechanisms such as green loans or subsidies since the late 2000s. The solar rollout is mainly driven by large commercial, medical and industrial institutions, as well as by relatively wealthy households who are able to invest several thousands of dollars in on-grid solar systems. As they rely extensively on expensive diesel generators to access electricity and have highly energy-intensive practices, these consumers are looking at solar energy to reduce their bills while securing their electricity supply. As in Bihar and Benin, the environmental argument is relatively marginal in comparison with the economic rationality and to overcome the intermittency of both the conventional grid and renewable energy, most solar energy systems are hybrid and connected either to the national grid or to local diesel generators. The non-regulatory policies implemented by public authorities lead to very varied quality of components and installation and to a very unequal and differentiated development according to the socio-economic conditions of users. Thus, the booming of the solar market, facilitated by the electricity crisis, is worsening inequalities, by widening the gap between grid-dependent consumers and users with multiple devices.

It must moreover be noted that, in Lebanon, the residential sector remains highly dependent on collective diesel mini-grids, which are often managed by private operators. These operators are strongly anchored in the urban landscape, hold small local-level monopolies (Verdeil, 2016) and are quite reluctant to integrate solar energy in their electricity mix. Similar dynamics are visible in Bihar, where diesel mini-grids are run by powerful informal entrepreneurs, whose role in preventing the deployment of solar solutions has been underlined by the existing literature (Szakonyi and Urpelainen, 2016). During interviews, these entrepreneurs would often declare not being opposed to transitioning from diesel to solar, but needing state support and incentives to do so, which were lacking due to the informal status of their businesses.

Implications for the future of the energy transition

The cases of Bihar, Benin and Lebanon illustrate how solar energy is often one of several options available to households and institutions to access reliable electricity supply. If in some cases solar can be used as a sole source of energy (as in the case of the households living in unconnected areas of Cotonou), in most places, solar energy is used in a complementary manner with other energy sources. Faced with weak or unaffordable conventional grids, consumers adopt complex supply strategies by combining solar panels, batteries, diesel generators and connections to mini-grids and the conventional grid. We argue that these strategies should not be seen as temporary, but rather as long-term processes. It is important to note that environmental motivations are rarely the main driver behind these strategies, which are mostly driven by issues of accessibility, cost, and suitability.

The deployment of solar energy therefore does not take place on a blank slate, but rather must be understood in relation to the practices and aspirations of various users and to the political and social structures that regulate both the conventional grid and other devices such as diesel generators. In other words, to use a term proposed by Sareen (Chapter 7), solar energy is ensconced in complex and context-dependent socio-technical landscapes, which it contributes to changing. Devices with which solar energy competes are often well embedded in day-to-day practices and controlled by actors with different logics and interests, such as the private diesel mini-grid entrepreneurs in Lebanon and Bihar, who hamper the development of solar energy to maintain their rent in supplying electricity. The limited development of PV testifies to the attachment of urban dwellers to the conventional grid, adapting their strategies and electrical configurations to the temporalities of the grid (deterioration, improvement, arrival, or expectation of its upcoming deployment). The limited development of renewables can also be explained by unequal resources of users.

Attention to these dynamics can help define regulations and subsidy programmes that promote the displacement of more polluting devices, while making solar energy solutions more attractive, more sustainable, and better adapted to various needs. As we have seen, markets in urban areas of the Global South already offer a wide array of solar devices, from the most basic solutions for the bottom of the pyramid markets, to large solar rooftops for well-off households or institutions. Yet, these markets are often under-regulated and existing subsidies are often ill-suited. Thus, for instance, the poorest urban dwellers often have to rely on solar devices which are neither subsidised nor covered by warranty. Low-cost unsubsidised solar solutions are indeed often better adapted to their needs and financial capacities, as they can be altered, repaired and upgraded over time (Balls, 2020). Promoting and subsidising modular solar kits, instead of standardised systems, could therefore be an interesting option (Balls, 2020). Similarly, one could imagine policies supporting collective solar solutions, which could be socially more equitable and more environmentally sustainable in terms of waste. Currently, solar energy is mostly being deployed on an individual basis in cities of the Global South, which furthers the social and political fragmentation that characterises these spaces. Collective solar mini-grids are emerging, but they are often run in urban areas on a profit basis, which limits their benefits in terms of social equity (Guillou and Girard, 2022). In Lebanon, a few private entrepreneurs are installing PV mini-grids in villages poorly served by the conventional grid (Chaplain, 2022). Their deployment in poorer territories, however, is constrained by a lack of public support. Solutions could also include the promotion of solar energy communities, which could make the investment in solar more affordable to households, or by offering subsidies to transition from diesel back-up to off-grid or hybrid solar rooftops for the commercial and industrial sectors.

In terms of governance, attention to actual energy access practices would first require quantifying these systems and producing official statistics. Second, it would entail involving municipalities and local authorities in energy planning to design place-based, adaptable and flexible systems. Municipalities and local authorities are uniquely positioned to integrate urban energy practices into public policymaking for solar deployment and to formulate locally relevant regulations, policies and subsidy programmes. Our fieldwork on the cognate aspect of heterogeneous electricity configurations unearths many challenges in this regard. In India, energy does not fall into municipalities' portfolio, and at the state and central levels, the transition to renewable energy is first and foremost imagined as the massive introduction of solar on the grid through large solar parks. Only in rare cases do local authorities show interest in energy issues (Basu, 2021). Conversely, in Lebanon, many municipalities would like to distribute renewable electricity to their inhabitants but are constrained by the lack of financial resources and by the legal framework. These mismatches in the scale of requisite action and the scale of authority and control over resources favours solar energy transitions that are largely fragmented, market-driven and socially inequitable in the complex overall picture of energy access in the urban and urbanising Global South.

Notes

1   Fieldwork in Lebanon was conducted by A. Chaplain, in Benin by M. Rateau and in Bihar by B. Girard, all thanks to funding by the ANR-funded Hybridelec project.

References

Bakker, Karen. 2010. *Privatizing Water: Governance failure and the world's urban water crisis*. Ithaca and London: Cornell University Press.

Balls, Jonathan N. 2020. Low-cost, adaptable solutions sell: Re-thinking off-grid solar diffusion at the bottom of the pyramid in India. *Energy Research & Social Science* 70: 101811. https://doi .org/10.1016/j.erss.2020.101811

Basu, Sumedha. 2021. Urbanizing India's energy transition. *Seminar* 744. Available at: https:// india-seminar.com/2021/744/744_sumedha_basu.htm

Björkman, Lisa. 2015. *Pipe Politics, Contested Waters: Embedded infrastructures of millenial Mumbai*. Durham, NC: Duke University Press.

de Bercegol, Rémi, Cavé, Jérémie, and Nguyen, Thai Huyen. 2017. Waste municipal service and informal recycling sector in fast-growing Asian cities: Co-existence, opposition or integration? *Resources* 6(4): 70.

Cavé, Jérémie. 2015. *La Ruée Vers L'ordure. Conflits dans les mines urbaines de déchets*. Rennes: Presses Universitaires de Rennes.

Chaplain, Alix. 2022. Strategies of power and the emergence of hybrid mini-grids in Lebanon. *Jadaliyya*. Available at: https://www.jadaliyya.com/Details/43932/Strategies-of-Power-and -the-Emergence-of-Hybrid-Mini-grids-in-Lebanon (accessed 6 June 2023).

Cross, Jamie. 2013. The 100th object: Solar lighting technology and humanitarian goods. *Journal of Material Culture* 18(4): 367–87.

Cross, Jamie, and Murray, Declan. 2018. The afterlives of solar power: Waste and repair off the grid in Kenya. *Energy Research and Social Science* 44: 100–109.

Guillou, Emmanuelle, and Girard, Bérénice. 2022. Mini-grids at the interface: The deployment of mini-grids in urbanizing localities of the Global South. *Journal of Urban Technology* 30(2): 151–70.

Lawhon, Mary, Nilsson, David, Silver, Jonathan, Ernstson, Henrik, and Lwasa, Shuaib. 2017. Thinking through heterogeneous infrastructure configurations. *Urban Studies* 55(4): 720–32.

Pilo, Francesca. 2022. Infrastructural heterogeneity: Energy transition, power relations and solidarity in Kingston, Jamaica. *Journal of Urban Technology* 30(2): 35–54.

Rateau, Mélanie, and Choplin, Armelle. 2022. Electrifying urban Africa: Energy access, city making and globalisation in Nigeria and Benin. *International Development Planning Review* 44(1): 1–26.

Rateau, Mélanie, and Jaglin, Sylvy. 2020. Co-production of access and hybridisation of configurations: A socio-technical approach to urban electricity in Cotonou and Ibadan. *International Journal of Urban Sustainable Development* 14(1): 180–95.

Siddiqui, Zakaria. 2018. 'Disempowerment of incumbent elite and governance: A case of Bihar's electricity sector'. In *Mapping Power: The political economy of electricity in India's states*, edited by Navroz K. Dubash, Sunila S. Kale and Ranjit Bharvirkar, pp. 50–71. New Delhi: Oxford University Press.

Smit, Warren. 2021. 'Urbanization in the global South'. In *Oxford Research Encyclopedia of Global Public Health*. Available at: https://doi.org/10.1093/acrefore/9780190632366.013.251

Szakonyi, David, and Urpelainen, Johannes. 2016. Solar power for street vendors? Problems with centralized charging stations in urban markets. *Habitat International* 53: 228–36.

Verdeil, Éric. 2016. 'Beirut: The metropolis of darkness and the politics of urban electricity grid'. In *Energy, Power and Protest on the Urban Grid: Geographies of the electric city*, edited by Andrés Luque-Ayala and Jonathan Silver, pp. 155–75. Oxon and New York: Routledge.

Verdeil, Éric, and Jaglin, Sylvy. 2023. Electrical hybridizations in cities of the South: From heterogeneity to new conceptualizations of energy transition. *Journal of Urban Technology* 30 (2): 1–10.

6

Accepting idealised solar farm portrayals? Exploring underlying contingencies

Harriet Smith, Karen Henwood and Nick Pidgeon

Introduction

This chapter considers visions as a cognate aspect of solar energy transitions, by examining how public acceptance of renewable industrial energy installations is shaped through solar visions at play across industrial and semi-industrial settings in South Wales in the UK. We question what work solar images and material installations do in relation to public messaging and cultural discoursing involved in the development of industrial-scale energy generation and provision for heavy industry. Industrial decarbonisation plans in the UK are organised through a clustering strategy, drawing together installations into geographic areas. The clusters have different problems and opportunities, as well as differing strategy agreements with research councils, corporations and government bodies. Thus, they operate regionally across sectoral networks and industry streams and provide localised solutions and visions. The clusters come together through the Industrial Decarbonisation Research and Innovation Centre[1] (IDRIC), which provides a meeting point and supports the development of key technologies, social science and policy frameworks. Our focus is situated within the South Wales Industrial Cluster[2] (SWIC) which is well-developed, drawing on good relationships between stakeholders, communities and the Welsh Government. The discussion is underpinned by data from deliberative workshops undertaken in 2022 at towns along the South Wales coast, home to key installations involved in a larger 'Celtic Freeport' plan.[3]

The chapter text is organised in the following way. We begin by arguing that photographs and visual media are important vectors for social change, including decarbonisation plans. Next, we provide an explanation of how visual media were developed as research tools and how we drew on a psychosocial framework in order to interpret images as lively actors that participants dynamically responded to. Then, we introduce the research areas, outlining key sociospatial aspects and industrial decarbonisation plans. Our argument is presented through the analysis of three images: first, a photograph of a solar farm in the UK which we interpret as idealising; second, a composite photograph of Pembroke Valero oil refinery with Hoplass solar farm, which we will call a situated image, that evokes complexly ideated responses; and third, we discuss a computer generated imagery (CGI) vision of Port Talbot dock, that includes a solar installation. Lastly, we draw out how solar visions are imbricated with public acceptance of the wider industrial decarbonisation and renewable energy plans. The chapter concludes by summarising what the different concatenations achieve in relation to fuller understanding of visualisations as a cognate aspect of solar energy transitions. Moreover, we demonstrate how solar infrastructure is utilised as a signifier of decarbonisation to frame public acceptance of industrial change.

Why images matter in public acceptance of renewable energy

More understanding and research are required to consider what is at stake in public acceptance of industrial changes to meet Net Zero. Energy is politicised (Bolsen, 2022), with unequal and unstable outcomes for locally impacted communities (Brock et al., 2021). It has been recognised that there is a danger of idealising renewable energies (Devine-Wright, 2022) and that this concern is closely aligned to concerns about how energy companies may be greenwashing (Bressand and Ekins, 2021; O'Neill, 2019). Further, Bressand and Ekins (2021) argue that the rollout of the first wave of renewables (solar, and to some extent land-based wind turbines) may have made acceptance of the next wave[4] of renewable energy technologies more problematic. They argue that changes may evoke 'deeper emotions' than integrated assessment energy models suggest (Bressand and Ekins, 2021, p.2). Furthermore, lessons from public acceptance in relation to food practices, mobility and Covid-19 pandemic protocols reveal gaps between policy and public opinion (Bressand and Ekins, 2021, p.2). When the public are left behind or ignored, things can go badly wrong (Pidgeon, Kasperson and Slovic, 2003; Pidgeon and Demski, 2012). Although researchers are interested in how knowledge interacts with power (Grubb and Wieners, 2020), there is a lack of academic attention to the import of decarbonisation images and visions (Biddau et al., 2022) in public messaging from stakeholders and as potent tools for achieving public engaged research. For the most part, attention to the use of images in media is from within journalism studies (O'Neill, 2019; Smith, 2017). While some scholars interested in public engagement and deliberative research have utilised visual tools, we consider a lack of attention to the publics' multilayered responses to images to be a missed opportunity in public acceptance research.

Visual dominance has expanded (circularly) through the increase in the use of images across cultures aligned to the digitisation of film and photography, information technologies media and social media as vectors for all communications. News media is the main knowledge transfer vehicle between industrial vision-makers, policymakers, government and the public (Hariman and Lucaites, 2007; O'Neill, 2019; Biddau et al., 2022). Consequently, as we will demonstrate, ocular-centric cultures are visually literate (Bamford, 2003). As a result, visual tools can convey knowledge democratically through engaged research designs that attend to participants' multiple complex readings (Harper, 2012). Visual literacy means the 'gaining of knowledge and experience about the workings of the visual media coupled with a heightened conscious awareness of those workings' (Bamford, 2003, p.1). This entails, for example, reading images in terms of narrative, identifying the subject, the citing syntax, the intention and considering the cultural context from where the image emanates. Recognising the affective tone of a photograph and being aware of how and why it is affecting, matters for decarbonisation, because images affect people in ways that shape their evaluation of risks (Pidgeon, Kasperson and Slovic, 2003) and public acceptance (Vespa et al., 2022). To quote Bamford (2003), 'The idea that "seeing is believing" is now a naïve concept. Manipulated images serve to re-code culture.' Meme cultures which circulate, repurpose and appropriate images across social media are an example of this. Working on the basis of publics as visually literate, psychosocial research has deployed visual narratives to temporally reframe, re-represent and re-imagine identities and, in this way, produce understanding of 'identificatory dynamics' (Henwood, Shirani and Groves, 2018, pp.10–11). Visual and narrative methods have also been used to promote reflection on intangible aspects of routine life where we 'lack the words to say what practice means' (Henwood, Shirani and Groves, 2018, p.1).

Visual analysis and psychosocial interpretation

This chapter draws on workshops we delivered by utilising visual materials produced for the two South Wales towns. We designed a range of tasks including a set of photographs that we asked participants to ascribe meanings and emotions to, and to select and assemble them into annotated constellations of meaningful objects and interrelations. We also presented expert visions and newspaper portrayals of related industrial plans for change. The localised photographs depicted a range of everyday life objects in the field research areas – for example shops, parks, fun, mess – added in alongside images of cross-temporal industrial infrastructures, for example an old gasworks, a hydrogen sea tanker and a solar farm. This invited an embedded cross-temporal and sector reading of industrial impacts that spoke to both personal local and global understandings (Hariman and Lucaites, 2007) and therefore invoked potentially personal and general meaning that grounded the future technology visions. We were able to understand how solar farms are ideated in comparison to other energy infrastructures as well as how solar farms are understood in place.

The analysis draws in part on Roland Barthes' (1977, 1982) analytics (e.g. this is a photographic representation of solar panels (signifier) and connotates a hopeful future). Theo van Leeuwen explains how objects within images can be understood as signifiers of shared connotations understood by Barthes as 'myths', through forming what Barthes called 'concatenations', meaning they deliver an overall connotation that is the sum of photograph objects in relation to one another, so '[i]t is their concatenation which connotes "myths", (van Leeuwen, 2004, p.9). Photographic objects include items such as solar panels, turbines and clouds, as well as stylistic and sensory objects such as light and texture. Here, we widened the definition of concatenation to operate across multiple photographs placed together into constellations of meaning. Our interpretation included the ways participants used the visual tools through felt engagement with the research objects (Blackman and Venn, 2010), enabling us to explore how affects shape experiences and opinions. Our interpretive framework opened up the possibility to produce reparative readings that express the ambiguous, contingent and uncertain ways (Sedgwick, 2003) in which people connect to visions of future technological change. We ideated the images on scales of liveliness according to how varied responses were to each given object and to how objects were ideated in differing constellations. In the following analysis and drawing concisely on this extensive basis, we illustrate how psychosocially engaged research provided a reparative vehicle to comprehend visualisations and images as a cognate aspect of solar energy transitions.

Reading solar landscapes

South Wales is the second biggest industrial and power carbon emitting region in the UK, due to the presence of heavy industry and power generation facilities. South Wales industry (10 mega-tonnes or Mt) and energy generation (6 Mt) together produce 16 Mt of emissions each year.[5] SWIC has developed a vision roadmap that includes electricity produced from solar farms and offshore wind and hydrogen produced from natural gas (blue hydrogen) and from offshore wind (green hydrogen). There are also plans to continue using fossil fuels by introducing carbon capture and utilisation and storage (CCUS) technologies. Carbon would be captured during industrial burning of fuels, stored and in part reused by suitable industries (such as some construction material and food production manufacturers). The majority of captured carbon is to be stored under the seabed of the North Sea. We selected research locations with the two largest emitters, the first workshop in Port Talbot as home to Tata Steelworks (the UK's largest steel fabricator) and the second workshop in Pembroke Dock with the Pembroke Valero Refinery nearby.

The South Wales coastal area is a ribbon of towns, beaches and industrial zones built up since the 1800s during the heyday of coal and related manufacturing. Pembrokeshire is home to several global gas and oil refineries. South Hook, Dragon LNG, and Valero are based along the mouth of a large estuary known as the Milford Haven Waterway. The towns share attributes associated with industrialisation and port infrastructures. However, they also differ in many respects. Port Talbot sits between hills and sea and has rooted communities connected via the steelworks and mining heritage. Port Talbot is also connected via a motorway to nearby cities, including the Welsh capital Cardiff. Pembroke Dock is situated on a peninsula and while the towns of Pembroke, Milford Haven and Neyland are nearby, they are small, with no motorway or railhead. Although also industrial, the oil refineries are disconnected from the area's history of fishing, boat building and as a military base. As the discussion unfolds below, it will become apparent how visions as a key cognate aspect of solar rollout are embedded in relation to public acceptance of wider industrial decarbonisation plans.

Solar to sunflower

The photograph of a solar and wind farm (Figure 6.1) is taken at standing level, accentuating the field grasses and solar panels that stand with light, apparently causing reflections from the bright blue sky and fluffy clouds – connotating synergy between panel and sky. The solar array rows recede into the horizon and centred in the background stands one wind turbine with a second also receding. The turbines draw the eye up to the white clouds and sky, and the array blends rhythmically with the field fauna through shadows falling from the rows of panels. The rhythm of receding technology with nature signifies time as an ongoing stable relation between the photograph object-signs. The concatenation (a combined object that connotes) is that the solar array is embedded into and synergised with the natural world, creating an idealised vision of future energy production.

The solar farm photograph was one of two non-local images that therefore held no directed place attachments for participants (Devine-Wright, 2022). Nor did they involve any locally recognised psychosocial commitments generated by life narrations, other lively ways of engaging with the historicising effects of socio-cultural discourses (Henwood et al., 2008), or more immediate, situated effects of meaning frames mobilised in social interactions (Henwood and Pidgeon, 2015). Instead, they spoke to imaginary framings and visions of change. Had this solar photograph been presented locally to where it was recorded, it would have become a situated photograph, but for our purposes the array is generalised and does not enact any placed associations.

We used the solar photograph in both town workshops. Each produced similar responses of being regarded – by most though not all participants – as clean and free. Although the two groups of townsfolk discussed different infrastructural change visions, overall, in both groups the image was ideated as hopeful and a representation of the future. Accepting this image, articulated by one participant as 'a view of the future', did not produce a situated acceptance in relation to thinking through what a specific solar farm would mean in place. Instead, the affectiveness of the image, active through texture, light and colour, culminates in an emplacement of the array set within a generalised natural scene. We regarded it as a static image that achieved a consensus of positive responses, often aligned with a close-up image of a sunflower, a running river and a garden. We interpreted these assemblage concatenations as expressing hope connected to cultural discourses of idealised living, positive wellbeing and health-orientated futures. We regard this form of acceptance as a surface acceptance in that there were

Figure 6.1 A solar and wind farm. © Soonthorn Wongsaita/
Shutterstock ID722347258

no localised signifiers. The concatenation of a hopeful future suggests
a form of acceptance that while connecting to an affectively active
emotional desire of hope, is different to acceptance grounded in material
commitment. We consider what work surface acceptance achieves after
presenting the situated solar farm photograph.

Solar replacing oil? Various readings of the Valero–Hoplass photograph

Around Pembroke Valero oil refinery are fields and scattered farmhouses.
Some of the local farms have changed from food production to solar, as
with the Hoplass Solar farm, visible in the mid-ground of the photograph
(Figure 6.2). The image was shot from a wide verge next to the only road
that weaves around the oil refinery and coast in this part of the peninsula.
On the horizon edge of the image to the left of the refinery, it is possible
to see a small amount of the Milford Haven Waterway, with South Hook
refinery visible on the right horizon line sitting on the Milford Haven town
side of the Waterway. In contrast to Figure 6.1, this is a landscape framing
with the solar farm visible at a distance. The image is a representation
of what anyone who takes the road will see, given the overcast weather
conditions that are common in this part of Wales. We consider the image

to be a 'mini-icon', since to local people, the Valero refinery holds a range of powerful meanings: memories of family employment, a current place of work, an 'eyesore', pollution, somewhere Just Stop Oil protests take place, and more.

Looking at the photograph objects reveals that the refinery towers stand surrounded by the cloudy sky and the solar panels hold the most light in the image. The texture of the field hedging both draws the eye in and accentuates a portrayal of the refinery and solar farm as a single installation standing together on the other side. The farmhouse is almost nestled in with the solar panels and the soft grey tones of the photograph further embed the solar farm into the industrial landscape. Whereas the turbines in the previous image join with the bright blue sky and clean fluffy clouds, in this image the refinery towers are embedded in an overcast sky. Had it been a sunny day, of course, things would have looked different. However, we consider that the overall concatenation would not be much changed. The solar array in this image, like the former idealised version, holds most light and works with the sky, connoted both by light and the refinery towers as guides upwards in place of the previous image's turbines. However, unlike the white wind turbines, the refinery towers signify heavy industry and fossil fuel. The overall muted tones and the distance of the shot produce a relation of grey sky, heavy industry and solar. The array from this framing appears less synergised with nature than in the idealised image and instead confers the sense of ground covered or wrapped in artificial materials, which in turn further connotates a relation between array and refinery.

Despite appearances, the two installations (oil refinery and solar farm) are not connected to one another, as far as we can tell.[6] As with the above description of the photograph, the material experience of seeing the refinery and solar farm is connective. Indeed, when we visited the site and made the photograph, we read the solar farm as connected to the refinery. This reading was also shared by some participants despite them living locally, who also read the solar farm as being part of the refinery. These participants suggested the photograph evidenced 'change' and 'contrasting energy sources for Valero', with one person wondering, 'Do you think they could knock down the refinery and put solar panels there?' Both the photograph and the material experience of being up close enact a relationship between solar and the fossil fuel refinery. An extension of this reading of the image is an optimistic ideation as evidence that solar is replacing oil, as the following quote illustrates:

that one's [Valero Photo] good because it shows that they're moving towards solar power, and that's like contrasting with how bad oil is and stuff. So it's like good and bad.

Figure 6.2 Pembroke Valero Oil refinery and Hoplass Solar Farm.
© The author

Linking solar to the oil refinery can thus be read as a hopeful ideation. However, it also suggests perhaps that solar is more of the same industrial spread across ground and nature, entwined with the polluting narrative of the refinery. While it appears hopeful (solar replacing oil), it also connotes with industry as an eyesore (the term 'eyesore' occurred multiple times with the Pembroke group) that threatens tourism, causes pollution and changes the farm industry. Moreover, as well as pollution, the Valero–Hoplass photograph was weighted in memory and conflicting feelings about job dependency and the future of the local economy. One participant emphasised how solar installations impact close-knit communities:

> This solar farm is like bordering his land, so the farmer who sold his land here, they've all like cut him off, fell out with him because he's a traitor.

Whereas the idealised solar photograph produced hopeful surface ideations, as well as hope of whether solar could replace oil, the Valero–Hoplass image also produced a combination of localised concerns and frictions such as a sense of treachery. A hopeful ideation of solar replacing oil more closely addresses the problem of fossil fuel emissions and new technological solutions. There is a distinction between the image that achieves surface acceptance as a 'view of the future' and the Valero–Hoplass photograph that evokes an assemblage of concerns as well as hopes. Providing an ambiguous image operated democratically within the workshop, as it provided wider vectors of meanings attached to localised knowledges and experiences, as well as broader conceptual considerations about land use and decarbonisation technologies. However, there is a problem in terms of acceptance based upon a myth that Hoplass Solar farm is owned by Valero and therefore marks a shift away from fossil fuel to clean energy. While we have no evidence to suggest that there is a relationship between Valero and Hoplass Solar farm, solar panels are increasingly being added to different industrial power generation facilities and installations.[7] During the workshops we asked townsfolk to consider wide-ranging technological changes including offshore wind, hydrogen production and CCUS infrastructures. Weighing up impacts of such introductions required people to think about what matters cross-temporally in their local areas. The portrayal that Valero is trying to move away from oil is concerning because it enacts a visual lie – albeit unintentionally – which set within the current context threatens future industrial changes.

Solar-superplace acceptance with hydrogen and CCUS?

Lastly, we discuss a CGI graphic from the marketing brochure for the Port Talbot Dock development, part of the larger strategy of the Celtic Freeport. The port is owned by the Association of British Ports, who have developed the plan in partnership with other Freeport bid members. We could not obtain timely permission to republish the CGI graphic, however we present a screen grab from Google Maps (Figure 6.3) which the CGI is developed from.

The CGI vision details how a solar array could be implanted into a 13-acre site on the Port Talbot Wharf between the boatyard and dinghy sailing club and industrial areas (marked 'A' on the figure). The plan aims to create a new area approximately retaining the dinghy club and the boatyard, but replacing what is currently scrub fauna. This CGI is part of a much larger proposal to transform and decarbonise the dock area, which envisions addressing the emissions from the Tata Steel plant (by utilising CCUS), developing the steel fabrications for off-shore wind turbines (to be towed out via Pembrokeshire to the Celtic Sea) and installing hydrogen and CCUS transportation infrastructure.

The CGI graphic plan is constructed from town spatial plans from Google Maps re-modelled to portray a vision of the future development that speaks stylistically to architecture and town planning, as well as computer games. The vision as it appears in the marketing brochure is disembodied from its place and requires contextualising to achieve public acceptance. The area holds deep memories for the townsfolk from days when the dock area was a large freight and coal transportation port and some historic buildings have already been lost – a source of sorrow in the town. The area is used by walkers, fishermen and people who keep boats. The steelworks is a contentious 'public-object' (Thomas et al., 2022), continually under threat of closure and having already seen loss of most jobs that the town used to rely upon.

We presented the CGI vision during a discussion of town future plans. We asked people if they agreed with the government concept of transforming industrial places into SuperPlaces (Business, Energy and Industrial Strategy, 2020). The development vision can be read as an imaginary of a SuperPlace with transformed industry, ready for the next generation of re-industrialised economy. Our workshop participants considered the likelihood for the SuperPlace transformation to manifest and weighed it with their experiences and opinions about the dock and port areas. 'Who's going to pay?' was a frequent response. They looked at the plans as people who know the area and regard it as 'home'. The

Figure 6.3 Port Talbot Dock. Source: Imagery ©2024 Google, Imagery ©2024 Airbus, Bluesky, Infoterra Ltd & COWI A/S, CNES/Airbus, Getmapping plc, Maxar Technologies, The GeoInformation Group, Map Data ©2024

CGI graphic illustrated a gap between (distant) vision-makers and people on the ground. The Port Talbot CGI visuals generated lively discussions, with some agreeing there is a good plan while others could not see how it would change their lives for the better, or perhaps worse. People were both wary of change and visually agile, considering what the vision creators intended them to see, as well as deliberating how to respond to the decarbonisation and development plans. They wanted to feel hope, but also saw a gap between idealisation and manifestation of what Barthes called 'myths' and what we call imaginaries and expert visions. Within the image, there is a concatenation that produces friction between the lived Port Talbot dock today and the future vision of a space-yet-to-come, designed for people that may or may not include current inhabitants.

We interpret solar in the CGI as an offer of a merging: a design that speaks to a bridge between town and industrial SuperPlace. The workshops made apparent that a conflation exists about the scale of energies required. Oftentimes, people suggested that solar panels on rooftops or razing the Valero oil refinery to extend the solar farms would be a viable, 'safe and free' solution for energy production. The Port Talbot CGI adds to such conflations in a similar way to the vista of Hoplass Solar farm doughnutting around the Valero refinery – visually connotating that solar will do more than it can.

Conclusion

There are similarities between how the solar farm portrayals enact with the other image objects, but there are also key differences. Images of solar installations may be (inadvertently/questionably) utilised to greenwash newer technologies (Bressand and Ekins, 2021) in order to garner public acceptance for hydrogen generation, storage and utilisation, as well as CCUS. Evidently, achieving Net Zero requires multiple energy installations, to power complex industrial infrastructure and the everyday needs of local communities. However, our findings raise concerns about what surface acceptance achieves through visions that may mask underlying issues and down the line engender public resistance and disruptions as decarbonisation takes off in the coming decades. But we also consider that hope is meaningful and that the idealised images of infrastructure were useful tools for affective expression that proved popular within workshop assemblages. The Valero–Hoplass photograph presented complex readings that evidence how people read images in thoughtful ways and how when presented with situated objects they are more likely to arrive at reparative positions. People weigh up realistic future visions in their own localities where they know what matters the most. In this way, situated images and visual tools can invite efficacy and democratising viewpoints alongside surface images that on their own can be regarded as curating idealised visions, thus inviting fissures between 'us' and 'them'. Attending to visions as a cognate aspect helps shed light on this vital dynamic in relation to solar energy transitions.

Notes

1   https://idric.org/about-us/
2   https://www.swic.cymru/
3   Freeports are part of the UK government drive to achieve 'levelling up'. At the time of conducting the research The Celtic Freeport, known also as a Green Freeport, was at the bid stage and the plans under discussion were uncertain. The bid has now been accepted. Each installation must hereafter gather environmental permissions and public support.
4   technologies such as green hydrogen and carbon capture reuse and storage(CCUS); see p.7.
5   See note 2.
6   Valero is a Texan-based oil and gas company. Valero has a second site nearby for fuel storage at an industrial complex that includes Dragon LNG, also with an associated solar farm.
7   We are unable to name any for ethical compliance reasons.

References

Bamford, Anne. 2003. The Visual Literacy White Paper, *Commissioned by Adobe Systems Pty Ltd.* Available at: https://aperture.org/wp-content/uploads/2013/05/visual-literacy-wp.pdf (accessed 10 June 2023).
Barthes, Roland. 1977. *Image, Music, Text*. London: Fontana.
Barthes, Roland. 1982. *Camera Lucida*. London: Paladin.
Biddau, Fulvio, Brondi, Sonia, and Francesco Cottone, Pablo. 2022. Unpacking the psychosocial dimension of decarbonization between change and stability: A systematic review in the social science literature. *Sustainability* 14: 5308. https://doi.org/10.3390/su14095308
Blackman, Lisa, and Venn, Couze. 2010. Affect. *Body & Society* 16(1): 7–28. https://doi.org/10.1177/1357034X09354769
Bolsen, Toby. 2022. Framing renewable energy. *Nature Energy* 7: 1003–4.
Bressand, Albert, and Ekins, Paul. 2021. How the decarbonisation discourse many lead to a reduced set of policy options for climate policies in Europe in the 2020s. *Energy Research & Social Science* 78: 102118. https:// doi.org/10.1016/j.erss.2021.102118
Brock, Andrea, Sovacool, Benjamin, and Hook, Andrew. 2021. Volatile photovoltaics: Green industrialization, sacrifice zones, and the political ecology of solar energy in Germany. *Annals of the American Association of Geographers* 111(6): 1756–78. https://doi.org/10.1080/24694452.2020.1856638
Business, Energy and Industrial Strategy. 2020. The energy white paper, powering our net zero future, UK Government. Available at: https://assets.publishing.service.gov.uk/government/uploads/system/uploads/attachment_data/file/945899/201216_BEIS_EWP_Command_Paper_Accessible.pdf (accessed 1 June 2023).
Devine-Wright, Patrick. 2022. Decarbonisation of industrial clusters: A place-based research agenda. *Energy Research & Social Science* 91: 102725.
Grubb, Michael, and Wieners, Claudia. 2020. Modelling myths: On the need for dynamic realism in DICE and other equilibrium models of global climate mitigation. *Institute for New Economic Thinking Working Paper Series* No. 112. https://doi.org/10.36687/inetwp112

Hariman, Robert, and Lucaites, John Louis. 2007. *No Caption Needed: Iconic photographs, public culture, and liberal democracy*. Chicago: University of Chicago Press.

Harper, Douglas. 2012. *Visual Sociology*. Oxon: Routledge.

Henwood, Karen, and Pidgeon, Nick. 2015. 'Gender, ethical voices and UK nuclear energy policy in the post-Fukushima era'. In *The Ethics of Nuclear Energy: Risk, justice and democracy in the post-Fukushima era,* edited by Behnam Taebi and Sabine Roeser, pp. 67–84. Cambridge: Cambridge University Press.

Henwood, Karen, Pidgeon, Nick, Sarre, Sophie, Simmons, Peter, and Smith, Noel. 2008. Risk, framing and everyday life: Methodological and ethical reflections from three sociocultural projects. *Health, Risk and Society* 10: 421–38.

Henwood, Karen, Shirani, Fiona, and Groves, Chris. 2018. 'Using photographs in interviews: When we lack the words to say what practice means'. In *The Sage Handbook of Qualitative Data Collection,* edited by Uwe Flick, pp. 599–614. London: Sage.

O'Neill, Saffron. 2019. More than meets the eye: A longitudinal analysis of climate change imagery in the print media. *Climatic Change* 163: 9–26. https://doi.org/10.1007/s10584-019-02504-8

Pidgeon, Nick, and Demski, Christina. 2012. From nuclear to renewable: Energy system transformation and public attitudes. *Bulletin of the Atomic Scientists* 68(4): 41–51.

Pidgeon, Nick, Kasperson, Roger, and Slovic, Paul. 2003. *The Social Amplification of Risk*. Cambridge: Cambridge University Press.

Sedgwick, Eve. (ed.) 2003. *Touching Feeling: Affect, pedagogy, performativity*. Durham, NC: Duke University Press.

Smith, Joe. 2017. Demanding stories: Television coverage of sustainability, climate change and material demand. *Philosophical Transactions: Mathematical, Physical and Engineering Sciences* 375(2095): 20160375. https://doi.org/10.1098/rsta.2016.0375

Thomas, Gareth, Cherry, Catherine, Groves, Chris, Henwood, Karen, Pidgeon, Nick, and Roberts, Erin. 2022. 'It's not a very certain future': Emotion and infrastructure change in an industrial town. *Geoforum* 132: 81–91.

van Leeuwen, Theo. 2004. 'Semiotics and iconography'. In *The Handbook of Visual Analysis*, edited by Theo van Leeuwen and Carey Jewitt, pp. 93–118. London: Sage. https://doi.org/10.4135/9780857020062

Vespa, Mariangela, Schweizer-Ries, Petra, Hildebrand, Jan, and Kortsch, Timo. 2022. Getting emotional or cognitive on social media? Analyzing renewable energy technologies in Instagram posts. *Energy Research & Social Science* 88: 102631.

7
Comparative visual ethnographies of the ensconcement of solar photovoltaics in the urban built environment of solar cities Jaipur and Lisbon

Siddharth Sareen

Introduction

This chapter draws on two comparative cases at the urban scale – Jaipur as one of the leading smart cities in India and the state capital of solar leader Rajasthan, and Lisbon as the European Green Capital 2020 and the capital of Portugal with high solar ambitions. Based on two months' fieldwork in Jaipur in 2016 and 2022 and six months' fieldwork in Lisbon 2017–2023, I draw out the diverse and contrasting ways in which solar photovoltaics (PV) are increasingly ensconced in the urban built environment. Differences within and across these iconic case contexts serve to illustrate the versatility of how this modular technology intermeshes with urban form, enforcement of building codes, local norms on use of spaces like rooftops and diffusing imaginaries of solar integration into street lamps, dustbins and fuel pumps to the point of ubiquity. These visual ethnographies are documented and interpreted with grounding in unfolding solar rollouts in both case cities and offer novel insights into a disregarded cognate aspect of solar energy transitions – namely the creative licence with which they unfold in entanglement with localised complexities and affordances. The range of observed ensconcement is discussed as an enabling insight towards more adaptive and permissive regulations to promote rapid urban solar rollouts, in keeping with situated urban practices, advancing scope for socio-technical transitions.

The urban geographies of solar photovoltaics in Jaipur and Lisbon

Walking along Jaipur's bustling streets, one often finds it difficult to cross major roads for minutes on end, with traffic careening past. Lisbon presents a different prospect in its cobblestoned inner streets, but head out to the suburbs north of the centre and pedestrians thin out. Cars and buses rule the day in the outskirts of the Portuguese capital, which is nonetheless quieter than the capital of the Indian state of Rajasthan. Yet, above the hustle of streets throbbing with urban life, elements of the landscape have begun to change since the late 2010s. Thinking back to impressions gained during months spent on fieldwork in Jaipur in 2016 and Lisbon in 2017–2019 and comparing them with impressions from comparative visual ethnography in both cities during 2022–2023 that feature directly in this chapter, some key similarities and differences come to mind.

Both cities are similar in having historically and culturally significant zones. Districts like Alfama in Lisbon are famous for their old buildings nestled into undulating hillsides and increasingly attract tourists (Fontes and Cordeiro, 2023). The market district by the Hawa Mahal in Jaipur is home to a bustling and polluted den of traditional trades and shops (Dadhich et al., 2018). Neighbourhoods with newer developments have more run-of-the-mill buildings and less of a cultural aspect, making each city a patchwork of different styles and character. Both cities embrace an emphasis on digital and technology leadership, with Jaipur among India's 'smart cities' and Lisbon having hosted the Web Summit from 2016 onwards. Both cities have a few high rise buildings interrupting a generally low urban skyline. Yet, Lisbon's hilly topography affords vertical relief, with many views over the cityscape and the Tagus river to its south from its wondrous *miradouros*, while some overpasses afford aerial perspectives over Jaipur. It is from these vantage points that differences become visible.

Over the past half a dozen years, these cities have begun to undergo an urban transformation up on their rooftops. Looking down on the roofs bathed in sunshine from one of many strategic perches in either bright city, it is increasingly common to see some roofs glinting back. The solar PV modules proliferating across richer households, and increasingly also across private- and public-sector enterprises, have multiplied in number and become characteristic constituents of the melange of urban rooftops. Yet, the patterns on display feature high contrasts that point to differences in the built environments, the building codes that govern interventions and the priorities that drive these solar rollouts. These aspects of such urban energy transitions are overlooked in thematic scholarship (Akrofi and Okitasari, 2022).

In this chapter, the embedded forms that solar PV assumes in these urban geographies are foregrounded in order to inform situated ways to approach and enable the expansion of solar PV in a range of variegated urban settings. The affordances of Jaipur and Lisbon and the socio-material co-shaping of the built environment and emerging rooftop solar PV geographies offer insight into policy and legislative measures that could yield a more organic and contextually appropriate set of spatial forms and configurations for transitions to urban solar energy for more localised clean energy usage.

Contrasting situated visual ethnographies of urban solar ensconcement

While enrolment has received conceptual treatment by geographers, this often relates to the agency and practices of actors or to institutional structures and processes (Bouzarovski and Haarstad, 2019). Embedment or embeddedness has a more spatial orientation (Hess, 2004). A more physically oriented term than either of these is ensconcement, which is expressive of an embrace, absorption, and evocative of a sense of melding together; distinct yet combined in close entanglement. It also offers a sense of sequence and an order of significance – one thing ensconces another, thus it in a sense precedes it and enmeshes it into the existing structure to evolve. This makes ensconcement an apt term – as yet not theorised in human geography, nor in energy social science – to capture the physical phenomenon encountered in comparative visual ethnography of Jaipur and Lisbon – the placement of solar PV modules and their accompanying paraphernalia into these evolving cityscapes.

In Jaipur, where building codes are less strictly observed and higher inequality among urban residents is evident in a wider variety of housing structures (ranging from affluent gated communities prioritised in urban planning to wayside shanties filling in neglected gaps), solar PV takes on a very wide range of appearances. Figure 7.1 shows how solar PV layers atop middle class housing in a suburb of Jaipur, considerably raising the height of the roofing with its mounting and displacing upward reaching branches of a tree to avoid shade on the panels. Endless such urban juxtapositions have cropped up around the city, changing its visual appearance in a clearly noticeable way at street level.

Figure 7.1 Solar PV atop middle class housing in a Jaipur suburb.
© The author

Figure 7.2 Solar PV integrated in Jaipur's street lighting above a shanty.
© The author

Figure 7.2, by contrast, shows a public use of solar PV on a street lamp, above a much poorer, improvised dwelling. This variety in spatial form and the built environment is typical of Indian cities (and many large cities in the Global South) that feature class societies with high variation in income levels and considerable informal housing. This is a stark reminder that ownership of solar PV is limited to relatively rich households, yet its beneficiaries and uses could potentially span a far greater range (see, e.g. Hadfield and Cook, 2019). The integration of solar PV into efficient street lighting is quite a common sight, as one local manifestation.

Many rooftops have canopies to offer shade against the harsh desert-like sunlight during evenings up on residential rooftops and terraces, which double up as places to dry laundry and recreate. Solar PV is often perched atop these canopies, or these are put up where once there were open rooftops. While strictly speaking this may well mark a breach of vertical building permits, the arguably temporary nature of such installations represents a discretionary space for households to act strategically and secure their local energy provision. In a detailed account of household practices and the urban built environment in Jaipur, Rosin (2001) offers culturally situated material analysis, although at the time understandably without attending to the role of solar PV, now an emergent consideration. On higher buildings, solar PV placements are barely visible from the street, but on two- and three-storey structures they often have high visibility and come to characterise a building anew.

A touch ironically, fuel pumps almost without exception feature solar PV, using the large expanses allocated to them to perch panels atop their roofs, visible to vehicles driving in off the street to fill petrol or diesel. Figure 7.3 shows one of many such examples, where the main building at the fuel pump has a solar PV array. Remarkably, even an emissions testing van parked by a toll station to conduct checks, on one of the main highways into Jaipur, was fitted out with panels on its rooftop. Figure 7.4 shows how solar PV is ensconced into the many adaptations to the rigours of weather in the arid local climate, here including a piece of clothing hung out the window for shade.

Larger commercial establishments with adequate rooftop space – which traditionally tend to be flat in local building styles – feature arrays of panels that are rarely prominent from streets and become visible at particular angles. Moving out of the city centre, one encounters solar PV linked to solar pumps for agriculture. As in Figure 7.5, this sight breaks up countryside fields, with distinctive arrays that farmers put up typically supported by quite generous public subsidies. Even so, diffusion has been slow, as farmers in this state periodically receive waivers on electricity

Figure 7.3 Solar PV is ubiquitous at fuelpumps in Jaipur. © The author

Figure 7.4 Solar PV on an emissions testing van off the highway in Jaipur. © The author

Figure 7.5 Solar PV for agricultural pumpsets on the outskirts of Jaipur. © The author

bill payments if they have fallen in arrears, often preceding elections, which is linked to being an important political constituency (Sareen and Kale, 2018). This creates perverse incentives not to adopt local renewable energy sources despite heavy subsidies of up to 90 percent when combining different public schemes. Heading out to quarrying and industrial estate sites at the urban periphery, solar PV becomes evident as a sound investment for these business and building owners with access to capital, who require cheap local electricity supply to lower their grid consumption – at relatively high industrial electricity tariffs – for steady everyday operational demand or to feed back as prosumers based on net-metering. Much of this solar PV capacity is mounted on factory and warehouse roofs as shown in Figure 7.6, with some of it integrated in new-builds which were coming up at various sites during field visits in 2022, Jaipur being a city that continues to grow.

In Lisbon, solar PV is much more emphasised in public messaging at the urban scale, and yet far less visible. The municipal energy agency Lisboa Enova has gamified solar PV in Lisbon as part of its Solis initiative (see https://www.solis-lisboa.pt), a centrepiece of the strategy that led to the city being awarded the status of European Green Capital during 2020 (Sareen and Grandin, 2020). Although 8 megawatts (MW) of solar

Figure 7.6 Roofs of industries at quarrying sites on Jaipur's outskirts with solar PV. © The author

capacity within city boundaries is the ambitious target, only a fraction of this has been accomplished, yet without certainty about the exact number, as Lisboa Enova is working to attract people to help map many small, unregistered urban installations. These peek out from rooftops in the distance when looking across the city's hillsides, as in Figure 7.7, yet disappear from view when at street level in proximity. The figure also shows that alongside solar PV, solar thermal is also in evidence, used to heat water. In stark contrast with Jaipur's canopies, rarely does one see any solar PV perched visibly atop a roof or jutting out as an overhang. The National Laboratory for Energy and Geology boasts a famous Solar XXI building on its sprawling suburban campus, which combines solar facades (dating back to when they were quite expensive for regular use outside of such a prestigious demonstration project) and energy efficiency interventions (Aelenei et al., 2019). Solar PV atop car parking lots has become relatively popular, with this application outside Solar XXI now seen at university parking lots elsewhere in the city. Yet, as evident in Figure 7.8, even while solar PV has begun to crop up on Lisbon's rooftops, a great deal of potential remains unexploited in this undulating urban enviroment.

Figure 7.7 Solar PV and solar thermal on some of the higher Lisbon rooftops. © The author

Figure 7.8 Some solar PV along with a lot of unused potential on rooftops in Lisbon. © The author

Figure 7.9 A solar PV panel resembling roof tiles in Lisbon at a solar developer's office. © The author

Figure 7.10 Neatly integrated limited solar PV atop middle class Lisbon buildings. © The author

Figure 7.11 A 'solar tree' at the national science museum in Lisbon's business district. © The author

This may well change in ways that retain continuity with the hilly city's prized appearance so linked to its rooftops, which are visible from various perspectives and galleries that look out over urban landscapes. As Figure 7.9 shows, the significance of layering on to existing built environments is not lost on solar developers, who are innovating solar PV panels that resemble these roofing tiles in colour and appearance. Plausibly, even with slightly lower efficiency, this aesthetic may enable more buy-in among potential adopters, not least in light of the electricity price crisis that hit households in Portugal hard in the early 2020s. But in terms of the rooftop solar PV that has actually been rolled out in Lisbon, it is rather affluent commercial or public institutional buildings as well as housing – such as in Figure 7.10 – that dominate, much of it located on relatively high buildings.

As a notable oddity, the headquarters of the national science museum in a newly developed part of town features a 'solar tree', with a white synthetic tree figure centred in a spectacular aerial ring next to the museum, covered with solar panels to popularise awareness of energy transitions. Depicted in Figure 7.11, this marks a striking intervention in the urban environment of the newly built business district in Lisbon's

Figure 7.12 Solar PV integrated into parking meters in Lisbon. © The author

eastern part. This museum works in engagement with cities across the country and is developing an offshoot centred on solar PV in the town of Moura further south in the Alentejo region. This cultural initiative links solar PV with a place that in 2008 hosted what was the largest solar energy park in the world. Its 36 MW size now pales in comparison to the gigawatt-scale solar parks enveloping large swathes of land, such as the Bhadla Solar Park in Rajasthan. The national science museum plans to showcase numerous aesthetic applications of solar PV in urban environments, such as panels fitted onto dustbins (to help compact waste), power mobile phone chargers next to public benches and such everyday uses. Solar PV is also taking hold in popular domains such as a local football stadium, where it has the prospect of becoming identified with one of three major football clubs that Lisbon hosts. And as with Jaipur's solar street lights, Figure 7.12 shows how solar PV is integrated into the urban environment in Lisbon, here used to power a parking meter.

Ensconcement as a cognate aspect of urban solar energy transitions

Both urban contexts show the versatility of how this modular technology intermeshes with urban form, enforcement of building codes, local norms on use of spaces like rooftops and diffusing imaginaries of solar integration into street lamps, parking meters and fuel pumps to the point of ubiquity. These visual ethnographies, documented and interpreted with grounding in unfolding solar rollouts in both case cities, offer novel insights into a disregarded cognate aspect of solar energy transitions, namely the creative licence with which they unfold in entanglement with localised complexities and affordances. This entangled unfolding is fittingly characterised as ensconcement.

Ensconcement, then, is key to situated solar energy transitions in urban contexts and can enable their acceleration as well as course correction if recognised and addressed through customised legislation and targeted incentives in urban energy policy. Attention to ensconcement is also *doable* at the neighbourhood and urban scales, by local actors such as municipal energy agencies and the actors they enrol through initiatives such as Solis. It is a point of departure for simple capacity building at the local scale that can recursively feed back into policymaking for more adaptive and rapid integration of solar PV into cityscapes.

Accounts such as that by Sovacool et al. (2022) point out numerous solar energy injustices at these local scales. Yet, how many of those could be overcome by informing policymakers at various governance levels, local decision-makers and doers about the sort of enablements that could help solar PV layer atop existing urban geographies, in equitable, situated ways? Bringing ensconcement into the toolkit that informs solar PV adoption strategies and backing it up with customised, place-specific governance interventions, can pave the way for accelerated and just small-scale solar PV rollouts in contexts as diverse as Jaipur and Lisbon.

Drawing on insights from both contexts, what can one learn about ensconcement that can inform policies in a wide range of cities looking to enable and expand urban solar rollouts? Articulated as a cognate aspect of solar energy transitions, ensconcement relates to the urban built environment as it exists within the grain of a given city. Solar PV unfolds in entangled relation with urban geographies even as the latter evolve through this co-shaping process, as residents identify possible ways to capitalise upon this clean, affordable source of local energy production within settings where spatial use is tightly competitive. Yet, the built environment exerts far more definitive influence over the emergence and expression of urban solar than vice versa. Solar rollouts cannot simply ride roughshod over existing land (and rooftop) uses in cities, despite unfortunate trends to this effect in more rural settings where marginalised groups' practices are often displaced by solar parks that embody large-scale and remote characteristics that have long been the norm in traditional centrally controlled energy systems. Instead, the emergent logics of ensconcement evident in both cases tell us something vital about the future forms that urban solar landscapes might take as they expand in each city. Indeed, these forms are essential to enable for renewable energy capacity additions to emerge in just, distributed ways, close to sites of energy consumption, yet respectful of existing everyday practices that they embed into and gently reshape towards reconfigured, democratic forms of material culture. A recent contribution by Lucchi et al. (2023) provides key design criteria, insights and terminology to this effect. Ensconcement has a purview that extends beyond their focus on historic built environments, however, lending itself to diverse contexts where solar PV can emerge.

In Jaipur, solar PV is increasingly a natural add-on to existing permissive spatial arrangements that stretch aesthetic and legal confines to realise pragmatic goals of local energy generation, with a view to affordable consumption in a context of growing demand. This is especially significant in the capital of the state of Rajasthan, which until the late 2010s constituted a context of a struggle to meet adequate and affordable energy access for basic electricity needs of households. Within a handful of years, it has become a state producing massive amounts of solar energy, much of it being transmitted out of state. Ensconcement also signals a responsibility to reflect the benefits of local developments such as solar PV rollouts for the inhabitants whose land hosts them and whose everyday practices they inflect through embedment and enrolment. Over time, manifesting such local benefits becomes a means to finding new ways for people to live with renewable energy: to rely on it, co-shape it and embrace it in welcome forms rather than develop animosity to it as an intervention that burdens them while benefiting populations and owners elsewhere.

In Lisbon, solar PV takes on more disciplined aesthetic forms, constructed in carefully curated conversations with building codes, urban targets and visions, and emergent popular imaginaries of urban energy futures shaped by cultural institutions. The efforts of a cultural institution such as the national science museum are significant in terms of championing solar PV as part of the Portuguese identity, much like cork or wine are the material focus areas of some of their other museums in various regions. There is scope to insert even more playfulness in how people relate with different forms of solar PV, as is beginning to take place as enabling legislation allows for solar energy communities to unfold. A natural next step would seem to be spatial logics of solar PV that transcend the constraint of single buildings and move into block- and neighbourhood-scale configurations for use, including production and consumption. Borràs et al. (2023) present insights that are promising in this respect.

Both contexts showcase powerful ways to advance and modulate the fitting of solar PV to accommodating (and adaptive) built environments, by linking its expansion to concepts that enjoy popular legitimacy. Visual ethnography of these variegated forms offer rich scope to inspire similar use cases in other settings where solar PV has not gained much traction yet, to avoid mistakes that spark backlash and to overcome policy inertia that stymies innovation and delays urban solar energy transitions.

In sum, the insights generated through attention to ensconcement here can inform action on accelerating urban solar energy transitions in coordination with the cognate aspect of the built environment these are necessarily overlaid on and become constitutive of. Such a situated approach to more adaptive and permissive regulations can merge agendas of solar rollouts and respectful urban development in aid of existing everyday needs and practices.

References

Aelenei, Daniel, Lopes, Rui A., Aelenei, Laura, and Gonçalves, Helder. 2019. Investigating the potential for energy flexibility in an office building with a vertical BIPV and a PV roof system. *Renewable Energy* 137: 189–97.

Akrofi, M.M., and Okitasari, Mahesti. 2022. Integrating solar energy considerations into urban planning for low carbon cities: A systematic review of the state-of-the-art. *Urban Governance* 2(1): 157–72.

Borràs, Irene M., Neves, Diana, and Gomes, Ricardo. 2023. Using urban building energy modeling data to assess energy communities' potential. *Energy and Buildings*: 112791.

Bouzarovski, Stefan, and Haarstad, Håvard. 2019. Rescaling low-carbon transformations: Towards a relational ontology. *Transactions of the Institute of British Geographers* 44(2): 256–69.

Dadhich, Ankita P., Goyal, Rohit, and Dadhich, Pran N. 2018. Assessment of spatio-temporal variations in air quality of Jaipur city, Rajasthan, India. *The Egyptian Journal of Remote Sensing and Space Science* 21(2): 173–81.

Fontes, Catarina, and Cordeiro, Graca Í. 2023. Portraying urban change in Alfama (Lisbon): How local socio-spatial practices shape heritage. *Urban Planning* 8(1): 110–20.

Hadfield, Paris, and Cook, Nicole. 2019. Financing the low-carbon city: Can local government leverage public finance to facilitate equitable decarbonisation? *Urban Policy and Research* 37(1): 13–29.

Hess, Martin. 2004. 'Spatial' relationships? Towards a reconceptualization of embeddedness. *Progress in Human Geography* 28(2): 165–86.

Lucchi, Elena, Baiani, Serena, and Altamura, Paola. 2023. Design criteria for the integration of active solar technologies in the historic built environment: Taxonomy of international recommendations. *Energy and Buildings* 278: 112651.

Rosin, R. Thomas. 2001. From garden suburb to olde city ward: A longitudinal study of social process and incremental architecture in Jaipur, India. *Journal of Material Culture* 6(2): 165–92.

Sareen, Siddharth, and Grandin, Jakob. 2020. European green capitals: Branding, spatial dislocation or catalysts for change? *Geografiska Annaler: Series B, Human Geography* 102(1): 101–17.

Sareen, Sareen, and Kale, Sunila S. 2018. Solar 'power': Socio-political dynamics of infrastructural development in two Western Indian states. *Energy Research & Social Science* 41: 270–78.

Sovacool, Benjamin K., Lacey-Barnacle, Max, Smith, Adrian, and Brisbois, Marie Claire. 2022. Towards improved solar energy justice: Exploring the complex inequities of household adoption of photovoltaic panels. *Energy Policy* 164: 112868.

8
Community solar struggles in Portugal
Abigail Martin

Introduction

Community solar projects are among a growing number of community energy projects through which non-traditional actors have come to participate in electricity markets, including residents, ratepayers, municipalities, cooperatives and more (Brisbois, 2019). These (typically) decentralised projects stand in contrast to the large, centralised, commercial energy producers that have historically dominated Western power sectors (Kelsey and Meckling, 2018). Whereas community energy programmes have been more prevalent in the European Union (EU), they have only recently begun to gather momentum in the United States (US). In both contexts, policy initiatives promise to expand this market segment further. This chapter considers dynamics in Portugal within these broader developments.

The concept remains quite new, with the implementation of support schemes underway. Member states have been incorporating new EU rules on community energy into national enabling frameworks. In the US, California, New York and other states have established strategies to incorporate community energy into their energy portfolios. More recently, the federal government has taken action under the US Inflation Reduction Act, which includes tax credits for developing clean energy in communities that host fossil fuel extraction, processing and generation facilities.

Overall, establishing community energy projects has been challenging. Many advocates have struggled to achieve their vision under new legal regimes. Moreover, projects have drawn the attention of large power sector incumbents and new corporate players, suggesting some degree of industry capture of a grassroots concept. In California, community-based economic development organisations and the environmental justice movement have struggled to implement their vision of community solar for well over a decade. In response to movement calls for supportive policy that would enable community-based and cooperative entities to produce solar power for local consumption and grid-injection, state policymakers established the Green Tariff Shared Renewables programme. Regulators implemented the programme in 2015 by directing large investor-owned utilities (IOUs) to establish a pool of renewable energy facilities between 500 kilowatts (kW) and 20 megawatts (MW) and offer premium 'green' electricity at higher rates. This is a far cry from what the social movement advocates proposed for community-conceived, community-owned-and-operated energy. In Portugal, which created an enabling framework for community energy in 2019, a similar phenomenon may be underway, whereby corporate entities are developing community-solar products and services without the need for community-led dissemination nor onerous processes of establishing cooperatives or other organisations.

To understand community solar struggles, this chapter examines advocacy and policy development for community solar in Portugal. I focus specifically on the role of social movements in shaping community solar initiatives. While growing energy social science research examines community solar developments as social enterprise and social innovation phenomena, there has been little attention to the role that social movements play in the development of community solar. Research for this case study draws upon qualitative research carried out between June 2022 and March 2023, including semi-structured remote video interviews on Zoom and in-person, participant observation, and document analysis.

Community solar

'Community solar' here encompasses initiatives that use photovoltaics (PV) to create 'community energy' – an umbrella term for a broad spectrum of energy-related initiatives, rather than a specific project size or organisational model (Creamer et al., 2018). Those who participate do so for various reasons, but in practice, projects are designed to accomplish only one of numerous possible outcomes, namely the democratisation of energy decision-making through greater participation in place-based community energy development; community or ownership of an energy system; improved social or economic status; and the sharing of collective benefits such as local energy security and energy cost savings (Bauwens et al., 2016; Koirala et al., 2016; Becker et al., 2017; Burke and Stephens, 2017; Creamer et al., 2018; van Veelen and van der Horst, 2018; Lazoroska et al., 2021). Many of these goals are tied closely to the climate justice movement, such as ensuring access to energy for all, producing it without harm to the environment or climate and rethinking overall attitudes and energy consumption (Szulecki, 2018). Community solar advocates add to this list of benefits the potential to strengthen local resilience, for instance improving a neighbourhood's ability to withstand power outages associated with storms or over-taxed grids (Faget, 2022).

Early community energy projects – whether solar, wind, or biomass – have emerged under diverse political, legal and financial support landscapes. There are numerous organisational forms, including cooperatives; non-profit organisations that manage energy production or supply for local residents; community development companies; 'transition' towns, re-municipalisation initiatives and eco-villages; social movements concerned with energy sustainability matters; and informal citizens' groups emphasising self-sufficiency and energy saving. These forms correspond to a broad spectrum of governing structures. Generally, community energy systems have functioned as jointly-owned distributed energy projects with decision-making structures designed to distribute benefits more fairly to non-corporate actors (Seyfang et al., 2013). Unlike conventional and commercial energy producers that are legally bound to profit shareholders, community energy projects have created innovations in resource ownership, energy production and consumption (Becker et al., 2017). Some scholars underscore the potentially transformative shifts that these projects may drive as they capture more market share (Stirling, 2014; Brisbois, 2019).

Community solar projects have attracted start-ups, utilities, energy conglomerates and other companies generally not considered 'community-members' or 'community-based' entities. These (more commercial) community solar offerings enable consumers to subscribe to shared energy systems, which are not necessarily owned, operated or developed by community actors. Thus, community solar projects encompass increasingly diverse actors on both sides of the electricity meter (see Figure 8.1). While in theory, community solar systems can be located on- or off-grid, in practice most are grid-tied and conceive of the (extra)economic benefits of community solar as being fundamentally created through access to the electricity market and grid infrastructure.

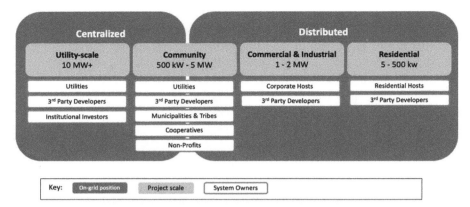

Figure 8.1 Community solar: scales and owners relative to other on-grid PV systems. Source: Authors' illustration based on literature review, herein Wainstein and Bumpus (2016) and Altunay, Bergek and Palm (2021). © The author

Cooperatives are the most established form of community solar in Europe (Horstink et al., 2020). Renewable energy cooperatives have various legal designs and operate as producers, suppliers, or both (Magnani and Osti, 2016). Generally, cooperatives operationalise a legal form for democratically owning and operating an energy project with shared value-creation for the cooperative's members or shareholders, with the majority being shareholders (Magnani and Osti, 2016). A common approach is to generate power for sale to the grid and distribute sales revenue to members of the cooperative, who may or may not physically belong to the same community. In addition, there may be payments to citizens who provide assets like rooftops, royalties or rents to the municipality, or discounts on energy bills (Hewitt et al., 2019). Currently, renewable energy cooperatives represent only a very small share of the electricity market and solar cooperatives even less so, due to the immense power and market share held by large energy companies. Cooperatives are often tied to or run by resident groups, municipalities and civil society organisations and typically struggle to secure political, technical and financial resources (Huybrechts and Mertens, 2014). In addition, community energy projects and cooperative approaches in particular have been neglected by energy policies that favour large-scale incumbent interests and do little to reduce legal and economic hurdles for small-scale solar, underscoring the need for supportive state interventions (Sareen and Nordholm, 2021).

However, growth in the community solar market segment is poised to strengthen with the entrance of for-profit entities with better access to finance compared to most cooperatives. Whether or not for-profit enterprises grow to dominate the community-solar market segment, their presence raises questions about whether and how other models of community solar projects are at risk of withering or, alternatively, have the potential to thrive and expand in a more crowded market for community solar.

Coopérnico and community solar in Portugal

The origins of community solar in Portugal

Coopérnico, an energy cooperative headquartered in Lisbon, has led advocacy for community solar in Portugal. The founding members formed the organisation in 2013 to democratise the energy sector with small PV systems for greater energy self-reliance in social establishments and create a 'social business'. Inspired and encouraged by European renewable energy cooperatives, they drew primarily upon ResCoop, a European network of 1,900 European renewable energy cooperatives and their 1,250,000 participating citizens. ResCoop promises to realise 'an economy and society based on co-operation rather than competition, within the boundaries of planet earth' and to advance energy efficiency, fossil fuel reduction, energy poverty alleviation and local economic activity (Friends of the Earth Europe, ResCoop.eu and Energy Cities, 2020). ResCoop supporters also claim cooperative producers reduce technology costs and maximise the potential sale of electricity to the grid, the profits of which are then used to lower electricity cost for individual families and invest in local development projects (Magnani and Osti, 2016).

As with many ResCoop initiatives, the founding of Coopérnico was led by an entrepreneur, Nuno Brito Jorge, who had experience in the energy sector at Energias de Portugal (EDP), ties to ResCoop, and access to social, cultural and economic capital. Inspired by his involvement with ResCoop activities in Brussels, Spain and France, Jorge gathered a group of founders, staff and almost 200 members within Coopérnico's first year (Otman, 2020). Jorge and other co-founders framed Coopérnico's mission not in terms of climate crisis or environmental values – as was often the tendency within ResCoop – but rather as a 'social business model'. Specifically, Coopérnico would offer the opportunity for ownership of energy production to reduce reliance on big producers, grow the share of solar energy in the Portuguese electricity mix and develop a social

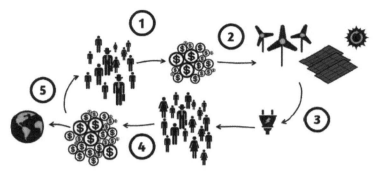

Figure 8.2 Coopérnico's vision for scaling community solar cooperatives. © Ana Rita Antunes

business model by bringing solar PV to public sector buildings and their cooperatives, eventually selling competitively-priced electricity to ratepayers and growing into a major electricity seller (Pinto, 2015). Figure 8.2 depicts this organisational vision.

Unlike other ResCoops, however, Coopérnico's social business model spoke to the discourse of socially-responsible 'impact investing' and 'green business' more so than that of grassroots social movements working on the climate crisis or cooperative-building initiatives. European renewable energy cooperatives often draw their political and organisational support from local civil society groups, professional environmental non-governmental organisations (ENGOs) like Friends of the Earth and sometimes supportive state actors (for instance, German cooperatives had supporters in the German Federal Ministry for the Environment, Nature Conservation and Nuclear Safety). Coopérnico began with ties to the environmental movement in Lisbon, but overall lacked civil society support to boost membership or achieve political gains through coalition-building. Portuguese environmental politics and social movement organising are a relatively young phenomenon. The large professional ENGOs have minimal presence, with comparatively low public involvement with environmental and other social movements, or support for environmentally-focused political parties compared to Germany and the Netherlands (Soromenho-Marques, 2002). Portuguese civil society has developed more awareness and capacity for environmental issues from the 2000s onward, but it has also become less radical over time, with most ENGOs joining the establishment, working closely with industry and government and adopting ecological modernisation discourses (Queirós, 2016).

This trend mirrors Portuguese civil society more broadly, whereby a brief but intensely active era of popular movements during the revolutionary period contributed to a more pluralistic and socially-inclusive society. In the ensuing decades, a range of participatory patterns for civil society took shape, ranging from partnership and consultation in the social welfare and neighbourhood movements, to weak consultation in the union movement (Fernandes and Branco, 2017). The victory of the moderates in November 1975 set the country on a path away from radical democratic practices and towards a pluralist representative democracy, with civil society less dense and less rich than other third wave democracies transitioning out of authoritarianism, like Spain (Rodrigues, 2015).

Coopérnico was also not bolstered by cooperative movements. Rural cooperatives grew substantially before the Estado Novo dictatorship, and many remain in operation today. After the 1974–75 social revolution, housing cooperatives grew substantially as well. Construction of cooperative housing peaked in 1989, with 4,582 dwellings built, before declining dramatically thereafter (Rodrigues and Fernandes, 2018). Although housing cooperatives generally enjoyed high levels of public and state support as a way to meet pressing public housing needs, many people today remember them suspiciously as a failed experiment that appeared to lack oversight, as evidenced by bankruptcies. The Portuguese public views agricultural cooperatives as prone to internal corruption (Castro and Moreira, 2022). However, structural forces severely undermined agricultural and housing cooperatives: policy for the cooperative sector was often ambiguous and contradictory (Namorado, 1999) and Portugal's entry into the Single Market and EU's liberal political regime dealt a major blow to the sector overall (Rodrigues and Fernandes, 2018). Thus, Coopérnico has not only lacked the support network provided by a strong cooperative movement, but also had to contend with general distrust of cooperatives rooted in collective memories.

Changes in the regulatory environment for community energy in Portugal

Initially, Coopérnico focused on building political capacity within the state and establishing its first energy cooperatives. In both realms, it received vital support from ResCoop, which helped finance its first cooperative projects and provided technical and policy advice, alongside the European Federation of Energy Citizen Cooperatives and Coopérnico's own fundraising efforts. Staff established important ties with parliamentary groups and other government actors to lobby for a supportive legal framework for energy cooperatives. Such efforts had little impact in an electricity policy arena dominated by large Portuguese and Spanish IOUs.[1] Moreover, government and institutional biases towards utility-scale solar farms translated to few supportive incentives, laws or regulations for small-scale solar (Sareen and Haarstad, 2021). For instance, the government did not create a legal definition for small-scale production until 2014, when it established Small Production Units for Self-consumption (UPACs in Portuguese) under the Portuguese Republic Decree Law (DL) 153/2014, which tied each production unit to a single meter. Moreover, although electricity could be injected into the grid, remuneration was limited to wholesale market prices (Camilo et al., 2017). This impeded the adoption of individual and shared solar systems to those with the disposable income and property assets as home-/building-owners. By 2018, Coopérnico had installed less than 1 MW, generated by small pilot projects located at single-meter properties in the wealthy Algarve region.[2]

Major policy changes arrived in 2019, when European directives on community energy demanded national transposition. Driven by pressure from ResCoop and advocates for inclusive, equal and efficient energy markets, the EU Renewables Directive (EU 2018/2001, 'RED II') and the Electricity Market Directive of 2019 (EMD) established basic definitions and requirements for enabling individual and collective self-consumption, renewable energy communities (RECs) and citizen energy communities (CECs) (European Commission, 2019). For individual and collective self-consumption, transposition should improve permitting procedures and incentives (e.g. discounts on grid chargers) that encourage prosumers to make use of their rooftops and enable those who do not own/occupy a standalone, detached house but are connected to the same distribution network to prosume, for instance those living in the same building. RECs (defined by RED II) refer to shared distributed renewable energy systems (i.e. solar PV, wind, hydro-power, geothermal, bio-energy) that have geographical proximity to the renewable energy source and go beyond profit maximisation to embody some form of inclusive, participatory decision-making on key issues (e.g. how to create and govern the system, how to distribute the electricity generated, how to distribute any financial returns). Despite diverse interpretations of the 'community' in how EU members transpose RECs into law, the overarching spirit of enabling frameworks for RECs is to broaden public participation in the energy sector for some geographical unit of community, as an alternative to conglomerate-controlled models of centralised electricity production (Hoicka et al., 2021).

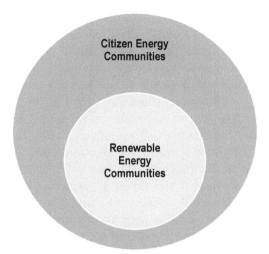

Figure 8.3 CECs and RECs. © The author

Similarly, CECs, defined by the 2019 EMD, are also legal entities with a non-commercial purpose that follow similar ownership and governance principles. However, CECs are more broadly interpreted than RECs, as these citizen energy communities lack geographical, technological and renewable restrictions associated with renewable energy communities. Transposition is meant to enable all three of these conceptions of community energy for distinct purposes, with overlap between RECs and CECs as shown in Figure 8.3. In addition, collective self-consumption can occur without creating an REC or CEC, and RECs/CECs offer benefits beyond collective self-consumption.

Portugal partially transposed RED II under legal directives DL 162/2019, DL 266/2020, and other corresponding reforms. The new legal regime made direct exchange between prosumers possible as either collective self-consumption arrangements or RECs, as long as it concerned bidirectional smart metering and similar voltage levels. Licensing and administrative requirements for collective self-consumption installations were also simplified to facilitate sharing with neighbouring buildings. However, the spatial limits for RECs were left vague, suggesting only that participants be located proximate to the installation (Campos et al., 2020).

Portugal's new regime has neglected other key elements of RED II – namely the establishment of a framework that will support and facilitate RECs. Without this, Coopérnico's vision for growing community solar cooperatives is limited. On the one hand, DL 162/2019 states that both collective self-consumption and RECs should receive renumeration for surplus energy supplied to the grid for the market value of that electricity and subsequent corresponding laws have reduced grid charges for both categories of community energy. On the other hand, the economics and administrative practices of creating and operating either category in Portugal remain challenging. Selling electricity back to the grid entails lengthy, complex applications and approvals and under the new regime, the grid operator (the National Directorate for Energy and Geology, or DGEG in Portuguese) has been overwhelmed by an increased volume of applications and assessments for collective self-consumption and RECs that wish to inject surplus into the grid.[3] Moreover, small collective self-consumption installations that sell surplus into the low-voltage distribution grid are required to form special management entities to oversee the project and obtain a contract with the supplier of last resort or another supplier on the free market, such as an aggregator under DL 162/2019. Remuneration is then calculated according to the average monthly market price of electricity in the Iberian Energy Market Operator, with a 10 percent injection penalty deducted (Camilo and Santos, 2023). RECs injecting into higher grid levels are not subjected to this tariff, but still face substantial red tape. The new regime legally allows RECs to produce, consume, store and sell their renewable energy within the REC, but to inject surplus into the grid as desired by Coopérnico, they must establish legal personhood, meaning to create a voluntary association of other legal persons such as local residents, small and medium-sized enterprises or local authorities. Once this formal legal entity exists, RECs face similar requirements as larger energy aggregators, suppliers and utilities seeking to market electricity, despite being small power producers.

Lawmakers also did not address CECs. To some extent, the recent passing of DL 15/2022 could inspire new CEC models. This is because it creates a new market in which energy suppliers bid competitively for 'Guarantees of Origin' to verify the delivery of energy to consumers that comes from a renewable source within Portugal – a market that could inspire new CEC models. However, auctions for Guarantees of Origin have stumbled and there still is no legal definition for CECs. As Ana Rita Antunes, one of Coopérnico's staff leaders, explains, Coopérnico has met the EU's standards to qualify as an energy community but cannot obtain recognition of this status due to the lack of implementing legislation in compliance with EU law, which prevents Coopérnico from gaining rights and privileges that will enable it to expand its deployment of community-based solar PV (Kern, 2022).

Growing community solar and social movements in Portugal

Bureaucratic hurdles and electricity prices have made it difficult to secure financial returns for community solar, which limits the kinds of projects Coopérnico pursues. Most Coopérnico cooperatives are collective self-consumption installations involving well-to-do condominium residents or property-holding social sector organisations. The people and organisations involved have the resources and value-systems conducive to seeking out, or responding positively to, cooperative solar proposals. These initiatives are also the result of Coopérnico's small staff working to recruit, organise and train those individuals and grassroots groups interested in building community solar systems; to educate and enrol local networks of homes, public buildings such as schools, kindergartens, social housing, community centres, non-profit elderly homes and other community-based organisations interested in making shared solar systems a reality through the use of their physical property; and to help usher projects through application processes. Coopérnico also provides financing and technical expertise once a project is operational. Still, Coopérnico staff remain rhetorically committed to the subsidiarity principle in project decision-making and the requirement that participants lead all aspects of project development and operations and to implementing inclusive democratic decision-making processes.

Coopérnico has guided the establishment of collective self-consumption projects for tenant-owned condominium properties of varying sizes. For instance, there is the small, single-building condominium in the upper-middle class neighbourhood of Oeiras[4] and the larger multi-building Condominio da Torre in Malha, Lisbon.[5] In addition, Coopérnico has established partnerships with social organisations (a daycare centre, a local aid organisation for people with disabilities) and municipalities. In this approach, Coopérnico members finance the installation, which supplies the organisation with power that reduces their electricity bills substantially, with around €1,000–5,000 in earnings per year for small projects that inject into the grid. As of mid-2023, Coopérnico had 3,944 members throughout the country, 2,859 contracts to install PV arrays and/or sell the electricity produced and had invested over €1,889,500 to finance community solar projects.

Coopérnico endeavours to graduate collective self-consumption cooperatives into RECs and eventually obtain CEC status for the cooperative as an energy supplier. The first REC in Portugal in the Vila Boa do Bispo parish began as a collective self-consumption cooperative in 2019, before expanding into an REC that includes several public buildings (the offices of the civil parish, a gymnasium, and a fire station), with plans to incorporate local condominiums and small local businesses. While there are a number of promising municipal-led initiatives under way that may achieve REC-status, Coopérnico's growth has been mainly through collective self-consumption, primarily involving well-resourced groups. These time-, resource- and knowledge-intensive endeavours require a 'core group' with the wherewithal to proactively lead project development.

There are other groups pursuing shared solar projects in Portugal, involving local politicians, municipal staff, university researchers and residents of multi-family apartment buildings – many with funding from the European level – that have also successfully developed collective self-consumption projects with PV. Even those that do not become cooperatives inevitably draw upon Coopérnico's work in some way, given that it has promoted the value of community energy for over a decade – in its entrepreneurial, advocacy and lobbying capacities – and has unique technical expertise on the Portuguese power sector's complex regulatory landscape at multiple levels of government. Still, in guiding Portugal's transposition of EU Directives, Coopérnico was ultimately unable to steer lawmakers towards greater appreciation of the benefits of supporting RECs.

Portugal's new regime has enabled more individual and collective self-consumption projects and lays the ground for a variety of new small-scale PV installations to emerge, including micro-grids and other peer-to-peer schemes, but hinders the expansion of RECs. The sector is simultaneously over-regulated and under-supportive of RECs. Portugal offers no feed-in tariffs (FITs) to support the small-scale projects in Portugal; despite the successful role FITs have played in promoting the development of solar energy, most member states, including Portugal, have phased them out and instituted more market-based and supply-side support strategies. Net-metering, which could support small-scale development, is also not allowed in Portugal. There are no high-level government targets, nor vision statements that outline community energy growth trajectories to help coordinate the sector and encourage more entrants. Neither are there incentives (financial or other, such as preferential grid access), nor industrial policy schemes designed to support RECs. It remains unclear whether the DGEG will identify these barriers and fulfil its assessment of whether RECs are a prospect available to lower income families, as required by the RED II transposition.

Notably, the new legal regime for collective self-consumption has attracted for-profit players into community energy. Cleanwatts, the self-titled 'climate tech' company based in Portugal is working to 'decentralize, digitalize and democratize access to clean energy' (Cleanwatts, 2023). The company claims its projects address energy poverty, target social welfare institutions and improve democratic decision-making over how benefits of solar will be scaled. Although Cleanwatts proclaims itself the 'leader in building and managing Renewable Energy Communities (REC)', it has not established an REC. Rather, it aims to support energy communities more generally, as demonstrated by its Cem Aldeias Social Energy Community project (Renováveis, 2022; We Electric, 2022). The project endeavours to train unemployed people to be domestic energy consultants, promising to impact 20,000 people in 100 small villages. It is unclear whether the Cleanwatts model will help increase access to the benefits of solar PV for those who lack the traditional assets needed to access community energy, or whether Cleanwatts' commercial gain will in some way undermine the realisation of co-benefits from community solar projects. There may also be concerns of 'community-washing' – a term used recently in Greece to describe a situation where private investors establish themselves as energy cooperatives to take advantage of favourable financing and regulations with impunity (Schockaert, 2021).

For Coopérnico, the challenge remains unlocking the potential of RECs and CECs, amid the dominance of existing centralised, corporate-dominated energy sector over regulators and policymakers. Elsewhere in Europe, community energy advocates and ResCoop projects have leveraged political mobilisations within the broader environmental and cooperative movements to help push politicians to improve support for decentralised energy. In Portugal, those traditions are not so common, yet it is telling that Coopérnico's largest groups of members are in Algarve, where communities of Dutch, Germans and other Europeans residents are already familiar with the cooperative movement.

To this end, although Coopérnico does not define itself as part of a social movement, its leaders have begun to explore how to strengthen and forge ties to other social movements, namely the environmental and climate movement, the housing rights movement and the 'transition movement' (which in Portugal refers to communities coming together to solve problems, develop mutual aid and build resilience). Coopérnico has increasingly re-framed community solar in terms of combatting energy poverty, bringing it closer to pro-poor housing movement groups like Habita!. Coopérnico is also drawing strength from a growing cooperative movement in Portugal, which enjoys strong youth support, especially in Lisbon. This younger generation is working to undo the older generation's negative perceptions and distrust of cooperative operations and to build a social movement. Though the agricultural sector still dominates cooperatives, new multi-sector cooperatives like Rizoma in Lisbon are gaining ground, with explicit orientations towards building community-based solutions to societal problems and commitment to participation in economic activities.

Conclusion

Coopérnico's role and experience in the burgeoning community energy segment of Portugal's power sector highlights the problems frequently associated not only with incumbent-driven energy policy (Brisbois, 2019) but more generally with decision-making processes controlled by those with more resources, subordinating democratic quality to economic growth. In Portugal, the predominant style of representative democracy depends on technocratic governance mechanisms, social individualisation and the globalisation not only of the markets but also of decision-making processes, underscoring the risk that poor participation poses for the quality of democratic governance (Smismans, 2006). Greater participation in representative democracies can help to expose economic inequalities, amplify the claims of those most dispossessed and strengthen the political power of those excluded from decision-making. Coopérnico's early focus on gaining access to decision-making forums and establishing influence with state actors through providing technical consultations and lobbying remains an important strategy for growing cooperative community solar projects. Yet, meaningful participation is not likely to come from influencing technocratic decision-making alone.

This chapter suggests that even as Coopérnico enjoyed greater access to the state, it could not successfully impart the value of its vision to decision-makers in the energy policy arena. Whether this outcome reflects regulatory capture by power sector incumbents, institutional inertia, a more mundane or less intentional blunder on the part of lawmakers, or some combination of these factors, the fact remains that the state incurs no legitimacy threats in failing to better support RECs, CECs and solar cooperatives.

Herein lies the value of social movements in just energy transitions. Ensuring meaningful participation in energy decision-making may require legitimacy threats. As sociologists have argued, social movements and civil society organisations (e.g. ENGOs) ascribe or deny legitimacy to political institutions and actors by developing 'repertoires of contention' – be it direct confrontation, mass media naming-and-shaming campaigns, protests, distributing leaflets, creating political coalitions, occupying buildings, or other direct actions (Tilly, 2004). The success of said repertoires depends centrally on the movement's strategic choices and structural limitations. If a movement's repertoire can demonstrate to the public that its claims are worthy, or that the state's actions or claims are illegitimate, it may gain public support, which can serve as a significant countervailing force to corporate interests in liberal democracies. Whether and how Coopérnico's exploratory relationships with social movements will contribute positively and substantively to the expansion of community solar remains to be seen. In Portugal and elsewhere, expanding community energy, and solar cooperatives in particular, may depend on charting new paths that widen democratic participation.

Notes

1   These include Energias de Portugal (EDP, previously Electricidade de Portugal, founded through the privatisation merger of 14 nationalised electricity companies), Galp, Gas Natural Fenosa, Endesa and Iberdrola Portugal.
2   For instance, the first Coopérnico projects included solar arrays located at Quinta do Caracol (a bed and breakfast) in the Algarve region, where there are a large number of vacation properties.
3   Before 2019, producers/UPAC of 200–1,500 watts of installed capacity could simply communicate their plans to the grid operator and DGEG, and only producers/UPACs of installed capacities exceeding 1,500 watts would need to obtain a license from DGEG. Under the new regime, projects as small as 30 kilowatts (kW) must now communicate their plans to DGEG and those exceeding 1 MW of installed capacity must obtain approval from DGEG.
4   The municipality of Oeiras lies on the outskirts of Lisbon and has the second-highest purchasing power and second-highest tax revenue among Portuguese municipalities after Lisbon; the highest growth domestic product in Portugal; and within Lisbon, it has the highest concentration of residents with higher education and the lowest unemployment rate (Safe Communities Portugal, 2022). Coopérnico approached residents of a six-apartment condominium about becoming members and starting a collective self-consumption project. Coopérnico guided the member-residents in creating the internal regulation, completing project registration and conducting monitoring and evaluation.
5   This project began independently in 2018, when the condominium association installed PV to power the building's common areas. Coopérnico helped expand the installation to 52 kW to power each apartment in the multi-building condominium.

References

Altunay, Maria, Bergek, Anna, and Palm, Alvar. 2021. Solar business model adoption by energy incumbents: The importance of strategic fit. *Environmental Innovation and Societal Transitions* 40: 501–20.
Antunes, Ana Rita. 2019 . Energia, poupança e custos. Colóquio E3S, Lisboa, 7 de novembro 2019.
Bauwens, Thomas, Gotchev, Boris, and Holstenkamp, Lars. 2016. What drives the development of community energy in Europe? The case of wind power cooperatives. *Energy Research & Social Science* 13: 136–47.

Becker, Sören, Kunze, Conrad, and Vancea, Mihaela. 2017. Community energy and social entrepreneurship: Addressing purpose, organisation and embeddedness of renewable energy projects. *Journal of Cleaner Production* 147: 25–36. https://doi.org/10.1016/j.jclepro.2017 .01.048

Brisbois, Marie Claire. 2019. Powershifts: A framework for assessing the growing impact of decentralized ownership of energy transitions on political decision-making. *Energy Research & Social Science* 50: 151–61.

Burke, Matthew J., and Stephens, Jennie C. 2017. Political power and renewable energy futures: A critical review. *Energy Research & Social Science* 35: 78–93. https://doi.org/10.1016/j.erss .2017.10.018

Camilo, Fernando M., Castro, Rui, Almeida, Maria E., and Fernao Pires, Victor. 2017. Economic assessment of residential PV systems with self-consumption and storage in Portugal. *Solar Energy* 150: 353–62.

Camilo, Fernando M., and Santos, Paulo. 2023. Technical-economic evaluation of residential wind and photovoltaic systems with self-consumption and storage systems in Portugal. *Energies* 16: 1805. https://doi.org/10.3390/en16041805

Campos, Ines, Guilherme, Pontes L., Marin-Gonzalez, Esther, Swantje, Gährs, Stephen, Hall, and Holstenkamp, Lars. 2020. Regulatory challenges and opportunities for collective renewable energy prosumers in the EU. *Energy Policy* 138: 111212.

Castro, Conceição, and Moreira, Tiago. 2022. Assessing corruption in agricultural cooperatives: Differences in the perceived level of corruption using microdata. *Studies in Business and Economics* 17(2). https://doi.org/10.2478/sbe-2022-0025

Cleanwatts. 2023. Available at: https://cleanwatts.energy/about-us/

Creamer, Emily, Eadson, Will, van Veelen, Bregje, Pinker, Annabel, Tingey, Margaret, Braunholtz-Speight, Tim, Markantoni, Marianna, Foden, Mike, and Lacey-Barnacle, Max. 2018. Community energy: Entanglements of community, state, and private sector. *Geography Compass* 12(7): e12378.

European Commission. 2019. Clean energy for all Europeans. Available at: https://doi.org/10.2833 /9937

European Parliament. 2018. Directive (EU) 2018/2001 of the European Parliament and of the Council of 11 December 2018 on the promotion of the use of energy from renewable sources (recast). Available at: http://data.europa.eu/eli/dir/2018/2001/oj

European Parliament. 2019. Directive (EU) 2019/944 of the European Parliament and of the Council of 5 June 2019 on common rules for the internal market for electricity and amending Directive 2012/27/EU. Available at: http://data.europa.eu/eli/dir/2019/944/oj

Faget, Jochen. 2022. Small cooperatives in Portugal produces solar energy. *Deutsche Welle*, 28 June. Available at: https://www.dw.com/en/why-small-cooperatives-in-portugal-produce-solar -energy/a-62275656

Fernandes, Tiago, and Branco, Rui. 2017. Long-term effects: Social revolution and civil society in Portugal, 1974–2010. *Comparative Politics* 49(3), Special Issue: Civil Society and Democracy in an Era of Inequality: 411–31.

Friends of the Earth Europe, REScoop.eu, Energy Cities. 2020. Community energy: A practical guide to reclaiming power. Available at: https://www.rescoop.eu/toolbox/community-energy -a-practical-guide-to-reclaiming-power

Hewitt, Richard J., Bradley, Nicholas, Compagnucci, Andrea B., Barlagne, Carla, Ceglarz, Andrzej, Cremades, Roger, McKeen, Margaret, Otto, Ilona M., and Slee, Bill. 2019. Social innovation in community energy in Europe: A review of the evidence. *Frontiers in Energy Research* 7: 31.

Hoicka, Christina E., Lowitzsch, Jens, Brisbois, Marie Claire, Kumar, Ankit, and Ramirez Camargo, Luis. 2021. Implementing a just renewable energy transition: Policy advice for transposing the new European rules for renewable energy communities. *Energy Policy* 156: 112435.

Horstink, Lanka, Wittmayer, Julia M., Ng, Kiat, Luz, Guilherme Pontes, Marín-González, Esther, Gährs, Swantje, Campos, Inês, Holstenkamp, Lars, Oxenaar, Sem, and Brown, Donal. 2020. Collective renewable energy prosumers and the promises of the energy union: Taking stock. *Energies* 13(2): 421.

Huybrechts, Benjamin, and Mertens, Sybille. 2014. The relevance of the cooperative model in the field of renewable energy in Europe. *Annals of Public and Cooperative Economics* 85(2): 193–212.

Kelsey, Nina, and Meckling, Jonas. 2018. Who wins in renewable energy? Evidence from Europe and the United States. *Research and Social Science.* 37: 65–73.

Kern, Ceilidh. 2022. Europe's citizen-led energy revolution tangled in legal difficulties. *Euractiv*. Available at: https://www.euractiv.com/section/energy/news/europes-citizen-led-energy -revolution-tangled-in-legal-fine-print/

Koirala, Binod P., Koliou, Elta, Friege, Jonas, Hakvoort, Rudi A., and Herder, Paulien M. 2016. Energetic communities for community energy: A review of key issues and trends shaping integrated community energy systems. *Renewable and Sustainable Energy Reviews* 56: 722–44.

Lazoroska, Daniela, Palm, Jenny, and Bergek, Anna. 2021. Perceptions of participation and the role of gender for the engagement in solar energy communities in Sweden. *Energy, Sustainability and Society* 11: 35. https://doi.org/10.1186/s13705-021-00312-6

Magnani, Natalia, and Osti, Giorgio. 2016. Does civil society matter? Challenges and strategies of grassroots initiatives in Italy's energy transition. *Energy Research & Social Science* 13: 148–57.

Namorado, Rui. 1999 'Cooperativismo e política em Portugal'. In *Cooperativismo, emprego e economia social* edited by Carlos Barros and J. C. Gomes Santos, pp. 96–115. Lisboa: Vulgata.

Otman, Abdul. 2020. Communities for Future: Coopérnico renewable energy cooperative, Portugal. Available at: https://communitiesforfuture.org/coopernico-renewable-energy-cooperative -portugal

Pinto, Mariana C. 2015. Boa Energia, the company that wants to reduce bills at home. *Publico*, 6 January. Available at: https://www.publico.pt/2015/01/06/p3/noticia/boa-energia -a-empresa-que-quer-diminuir-as-contas-la-de-casa-1822297

Queirós, Margarida. 2016. 'Environmental knowledge and politics in Portugal: From resistance to incorporation.' In 'Environmental Knowledge, Environmental Politics', edited by Jonathan Clapperton and Liza Piper, *RCC Perspectives: Transformations in Environment and Society* 4: 61–68. https://doi.org/10.5282/rcc/7701

Renováveis Magazine. 2022. Projeto '100 aldeias' da Cleanwatts ultrepassou objective. *Renováveis*, 16 November. Available at: https://www.renovaveismagazine.pt/projeto-100-aldeias-da -cleanwatts-ultrapassou-objetivo/

Rodrigues, Camila. 2015. Participation and the quality of democracy in Portugal/A participação e a qualidade da Democracia em Portugal. *Revista Crítica de Ciências Socias* 108: 75–94. https:// doi.org/ 10.4000/rccs.6111

Rodrigues, Camila, and Fernandes, Tiago. 2018. The cooperative movement in Portugal beyond the Revolution: Housing cooperatives between shifting tides. *Portuguese Studies* 34(1): 52–69.

Safe Communities Portugal. 2022. PSP Lisbon Outer Divisions. Available at: https://www .safecommunitiesportugal.com/psp-lisbon-outer-divisions/# (accessed 29 June 2023).

Sareen, Sareen, and Haarstad, Håvard. 2021. Decision-making and scalar biases in solar photovoltaics roll-out. *Current Opinion in Environmental Sustainability* 51: 24–29.

Sareen, Sareen, and Nordholm, Amber J. 2021. Sustainable development goal interactions for a just transition: Multi-scalar solar energy rollout in Portugal. *Energy Sources, Part B: Economics, Planning, and Policy*. https://doi.org/10.1080/15567249.2021.1922547

Schockaert, Heleen. 2021. Greek energy communities at risk: Urgent action needed. REScoop. eu. Available at: https://www.rescoop.eu/news-and-events/press/development-of-energy -communities-in-greece-challenges-and-recommendations (accessed 26 January 2024).

Seyfang, Gill, Park, Jung J., and Smith, Adrian. 2013. A thousand flowers blooming? An examination of community energy in the UK. *Energy Policy* 61: 977–89.

Smismans, Stijn. 2006. 'Civil society and European governance: From concepts to research agenda'. In *Civil Society and Legitimate European Governance,* edited by Stijn Smismans, pp. 3–19. Cheltenham: Edward Elgar Publishing.

Soromenho-Marques, Viriato. 2002. 'The Portuguese environmental movement'. In *Environmental Activism in Society,* edited by L. Vasconcelos and I. Baptistapp, pp. 85–127. Lisbon: Luso-American Foundation.

Stirling, Andy. 2014. Transforming power: Social science and the politics of energy choices. *Energy Research & Social Science* 1: 83–95.

Szulecki, Kacper. 2018. Conceptualizing energy democracy. *Environmental Politic* 27(1): 21–41.

Tilly, Charles. 2004. *Contention and democracy in Europe*, 1650–2000. Cambridge: Cambridge University Press.

van Veelen, Bregje, and van der Horst, Dan. 2018. What is energy democracy? Connecting social science energy research and political theory. *Energy Research & Social Science* 46: 19–28.

Wainstein, Martin E., and Bumpus, Adam G. 2016. Business models as drivers of the low carbon power system transition: A multi-level perspective. *Journal of Cleaner Production* 126: 572–85.

We Electric. 2022. Cleanwatts e Go Parity querem alargar projeto '100 Aldeias'. We Electric, 14 September. Available at: https://welectric.pt/2022/09/14/cleanwatts-e-go-parity-querem -alargar-projeto-100-aldeias/

9

Geopolitical ecologies and gendered energy injustices for solar power in Ghana

Ryan Stock

Introduction

Ghana's renewable energy transition is rapidly underway. Until recently, the vast expanse of solar arrays that now characterises the communities of Kaleo and Lawra was forest commons. From shea to firewood, women from resource-dependent households would enter these spaces daily to retrieve resources essential to household reproduction. Utility-scale solar power in Ghana presents a paradox of development. Although erected by foreign firms with globalised capital to mitigate climate change and provide much-needed electricity to the Upper West region, solar infrastructures have eviscerated the lifeways and livelihoods of many local residents, exacerbating their economic and energy poverty. This chapter illuminates global energy geopolitics and local gendered injustices as cognate aspects of solar development in Ghana, drawing on mixed-methods fieldwork conducted between September 2021 and February 2022.

The Upper West is among Ghana's poorest regions, with 66 per cent of residents suffering from multidimensional poverty and rampant unemployment (Ghana Statistical Service, 2020). The region also experiences disproportionate energy poverty, with 98.2 per cent of the households reliant on traditional cooking energy sources (Adusah-Poku and Takeuchi, 2019), combined with a substantial lack of connectivity to the national grid. Beleaguered by crippling economic and energy poverty, residents of Kaleo and Lawra are also disproportionately vulnerable to climate impacts. Erratic rainfall events of recent years simultaneously

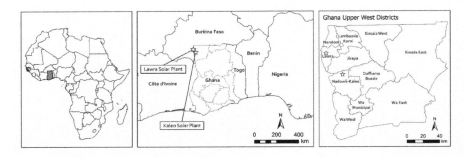

Figure 9.1 Location of the Kaleo Lawra Solar Plant in Upper West, Ghana. © The author

threaten residents with droughts and floods (Nyantakyi-Frimpong, 2020). Climate risks to the region are intensifying and vulnerable households are having difficulty adapting (Bezner Kerr et al., 2022).

Despite greenhouse gas emissions remaining at very low per capita levels beyond reproach, the vulnerable nation has pledged to 'scale-up renewable energy penetration by 10% by 2030' (MESTI, 2021, p. 26), to meet Paris Agreement commitments of reducing greenhouse gas emissions by 45 per cent (Government of Ghana, 2015). As outlined in the National Climate Change Master Plan, Ghana will '[p]romote the production and use of zero-carbon emissions sources of renewable energy (solar, wind, geothermal and mini-hydro)' (MESTI, 2015, unpaginated). In the next few years, Ghana will develop 150–250 MW of solar power. While renewable energy infrastructures like the Kaleo Lawra Solar Plant are being developed throughout the nation, most of Ghana's electricity is still derived from thermal plants powered by natural gas, diesel or crude oil (61 per cent), with a sizeable amount of power being derived from hydroelectric dams (38 per cent). Throughout the nation, the average rate of household electricity access is 83.24 per cent (Ministry of Energy, 2022). However, connectivity is much lower in rural areas. Much to the chagrin of energy-poor regions like the Upper West, Ghana exports excess electricity to Benin, Burkina Faso and Togo (Adusah-Poku and Takeuchi, 2019).

Solar energy has a bright future in Ghana. The nation has exceptional solar irradiance, conducive for generating electricity through solar power plants. In 2020, the state electrical utility, Volta River Authority (VRA), acquired 44.92 hectares in Kaleo and 6.13 hectares of land in Lawra to develop the 35-megawatt Kaleo Lawra Solar Plant, which would be split between the two locations (Figure 9.1) and was estimated to provide electricity for 32,200 households (VRA, 2022). The majority of lands enclosed for the solar plants were public – spaces that were frequently accessed and productively used by adjacent peasants. The VRA held 'extensive consultations' with land-owning families and regional land bureaucracies, eventually acquiring the private lands of four landowners on 50-year leases. Landowners were paid a 'negotiated sum' in full prior to project development, an amount based on 'arm's length discussions on current market values within the neighbourhood' (VRA, 2020a, pp.4–11). The VRA also held a stakeholder hearing and public consultation in 2012 and a pre-construction public hearing in 2017 during the environmental impact assessment process (VRA, 2020a). Both solar sites are presently operational, with the Lawra site (6.85 MW) being commissioned by President Nana Akufo-Addo on 10 October 2020 and the Kaleo site (13.8 MW) being commissioned on 23 August 2022 (Nartey, 2022). The Kaleo Lawra Solar Plant was built to concomitantly combat climate change, economic and energy poverty in the region. Yet, as this study will reveal by channelling voices of impacted stakeholders, the global circulation and accumulation of solar capital has produced localised energy dispossessions through solar enclosures that exacerbate social inequalities on the ground. The Kaleo Lawra Solar Plant is a generative case study through which to explore the cognate aspects of energy geopolitics and gendered injustices from solar enclosures.

Geopolitical ecology and solar energy

Nations and firms of the Global North are presently investing in and elaborating massive renewable energy infrastructures in the Global South to mitigate climate change and ensure energy security. Consider this recent statement from the government of Germany:

> There is hardly a continent on the planet that has more renewable energy resources than Africa. With thousands of hours of sunshine per year, solar power is an obvious option for Africa… From the smallest solar power system to state-of-the-art power plants: Africa's future is green. (BMZ, 2021, p.4)

With foreign assistance from the Global North, Ghana is rapidly building out renewable energy infrastructures. The Kaleo Lawra Solar Plant is a prime example, built under the auspices of achieving Sustainable Development Goal 7 (affordable and clean energy) by 2030 (VRA, 2020b) and net-zero greenhouse gas emissions by 2050 (IPCC, 2022).

The Kaleo Lawra Solar Plant must be contextualised within larger geopolitical machinations occurring through foreign assistance for climate action. In recent years, there has emerged a German–Ghanaian partnership that has accelerated Ghana's climate mitigation efforts and renewable energy transition. According to Germany's Federal Ministry for Economic Cooperation and Development, 'Africa's renewable energy potential must be utilised and the continent must pay a bigger role in the global green energy revolution' (BMZ, 2021, p.21). Germany's Marshall Plan with Africa has established the priority sectors for Ghana in '[r]enewable energy and energy efficiency', and mobilises private investments for infrastructures and employment opportunities (BMZ, 2021). Since 2017, the German government has supported Ghana in the financing and implementation of renewable energy infrastructures. Kreditanstalt Für Wiederaufbau (KfW), the German development bank, funded the Kaleo Lawra Solar Plant through an affordable loan agreement between the Ghanaian and German governments in the amount of €22.8 million (VRA, 2022). The Kaleo Lawra Solar Plant must also be situated within the larger G20 Compact with Africa (CwA), forged by Germany to promote private investments in recipient nations, with the goal of '[l]everaging private capital for infrastructure projects' in Ghana (CwA, 2017), such as solar power plants. The CwA highlights Ghana's recent efforts to '[f]ocus on renewable energy and energy efficiency' in the energy sector, seeking private capital to '[i]ncrease penetration of renewable energy in the energy mix, to achieve at least 10% of the generation mix by 2020' as an investment opportunity (CwA, 2017). To achieve these goals, the Ghanaian government has received financial and technical assistance from many firms and governments of the Global North.

In 2019, Elecnor SA, a Spanish firm, partnered with Volta River Authority to build the Kaleo Lawra Solar Plant. In addition to solar arrays, Elecnor installed control buildings, switching lines, medium voltage transmission lines, inverters, and transformers at each solar site (Elecnor, 2023). Elecnor Group specialises in energy infrastructure, boasting an extensive portfolio of renewable energy projects throughout the global South. Each phase of the Kaleo Lawra solar project involved Tractebel Engineering, a German firm that served as project consultants for VRA alongside Elecnor, in charge of project design, management, implementation, contracting and adherence to environmental and safety standards (Tractebel, 2022). Tractebel ostensibly elaborated the project with the objective of supporting 'the VRA in achieving its goal of becoming a renewable energy player in the region' (Tractebel, 2022). The Kaleo Lawra solar project was not Tractebel's first in the nation. In 2013,

Tractebel also served as project consultant for the 2.5 MW Navrongo Solar Power Station in the Upper East region. Given an evergreen Germany–Ghana partnership, it likely will not be its last.

Northern commitments to climate action in the Global South are largely in response to pollution and ecological degradation caused by centuries of imperial capitalist accumulation and industrial development (Moore, 2017), rendering nations like Ghana disproportionately vulnerable to climate impacts and ripe for geopolitical manoeuvres through mitigation infrastructures like solar plants. Energy system transformations towards decarbonisation are shifting energy geopolitics and renewing a focus on global energy security (Bradshaw, 2009). Energy security is of particular concern for Ghana. Yet, despite a reasonably efficient and reliable track record for a West African utility, the International Monetary Fund recently intervened to separate the Volta River Authority and restructure their debt (IMF, 2016), exacerbating local energy insecurities and causing power outages. In the wake of neoliberal restructuring, state and non-state actors of the Global North vie for power (exercised through solar power) in a nation grappling with economic and energy poverty (Adusah-Poku and Takeuchi, 2019).

The hegemony of the Global North in a carbon-constrained world is partly sustained through neocolonial production relations in renewable energy infrastructures in the Global South. For example, utility-scale solar parks in India that receive globalised financial capital (re)produce racialised and gendered vulnerability of smallholding peasants through land and energy dispossessions, as solar enclosures dislocate resource-dependent lower-caste women from access to firewood and other non-timber forest products (Stock and Birkenholtz, 2020). Similar patterns of dispossession and injustice occur globally with solar infrastructures rolling out at breakneck celerity and colossal scale (Sareen, 2022), a historically unrivaled energy transition for the Global South with grave local consequences. The development of solar micro-grids in West Africa are also beset by distributional inequalities of benefits and burdens that characterise utility-scale solar (Cantoni et al., 2022), notably the conflagration of land-use conflicts and perpetration of resource dispossession in Ghana (Nuru et al., 2022). As this study will reveal, the Kaleo Lawra Solar Plant falls within a global genealogy of geopolitically strategic solar interventions that (re)produce the social vulnerability of marginalised populations in the Upper West, climate action afforded through gendered energy and resource dispossessions.

This chapter emphasises energy geopolitics and gendered injustices as cognate aspects of solar transitions. Scholars of energy geopolitics are increasingly critical of (neo)colonial and imperial motivations and modalities of influence flowing from the Global North to the Global South in response to the climate crisis (Dalby, 2014). Advancing these concerns, Bigger and Neimark (2017, p.14) introduce the concept of *geopolitical ecology*, loosely defined as a framework that 'combines the strengths of political ecology with those of geopolitics in order to account for, and gain a deeper understanding of, the role of large geopolitical institutions, like the US military, in environmental change'. In this study, I illuminate local energy injustices by tracing the contours of geopolitical gambits that influence solar energy development in Ghana. Data for this study was collected via mixed methods fieldwork occurring between September 2021 to February 2022 in Kaleo and Lawra.

Gendered solar energy injustices

The Kaleo Lawra Solar Plant is widely considered a powerful example of climate multilateralism, insofar as nations of the global North marshalled resources and expertise to initiate and implement a successful public–private and North–South partnership with VRA to develop the solar plant which is now operational. However, global energy geopolitics have manifested local energy injustices. Although an indisputably strategic energy infrastructure in Ghana's trajectory towards decarbonisation in an effort to concomitantly combat climate change and energy poverty, the development of the Kaleo Lawra Solar Plant has ironically exacerbated the economic and energy poverty of local people. In the words of a Dagaaba animist practitioner from Lawra, 'The area has been infested with solar panels' (interviewed on 26 February 2022). After acquiring the land, the VRA cleared these spaces of all trees and shrubs and levelled the ground.

Enclosing vast swathes of land to host the solar arrays dislocated resource-dependent households from access to firewood and economic trees, such as shea and dawadawa. Despite being public lands, these forest commons were accessed daily by women retrieving vital resources necessary for household reproduction. According to a Dagaaba Muslim woman from Kaleo:

> It is the shea trees that have been cut down, which impacted women because they pick shea fruits and nuts and women in the north pick cocoa. Also, some depend on firewood and charcoal for

their survival. But now, where they used to harvest these has been cleared. They can no longer have it unless they are far into the bush. (interviewed on 22 February 2022)

Given the inadequate electrification of households in Kaleo and Lawra and biomass as a central fuel source, firewood resource dislocations have exacerbated the energy poverty of adjacent households. Respondents from both solar sites reported that solar enclosures impacted access to firewood. In addition to consuming products retrieved from the forest, some households processed these resources and derived their livelihoods from so-called economic trees that provide (*inter alia*) fruits, nuts and medicinal compounds. A Christian Dagaaba woman from Kaleo expressed it succinctly: 'Those people, if they don't have any other livelihood source, will be hungry' (interviewed on 24 February 2022). Solar enclosures have exacerbated the economic poverty of many adjacent households, who now must purchase most food they consume and obtain essential resources for their livelihoods from the market.

Aggravating the deprivations faced, the solar plant has not provided employment opportunities for local people. In contrast to employment guarantees by developers, locals were only hired temporarily to clear and level land and for construction of the solar plant, jobs which have since vanished since the project has been commissioned. Comprising the lion's share of the solar labour force at both sites are highly credentialed non-local men. The dearth of solar employment is of particular concern to many households who relied on farming the recently fenced lands, positioning them between a rock and a hard place. In the words of a Christian Dagaaba man from Lawra, 'The land used for the project is no longer suitable for agriculture. People laboured on the project to make a wage so that they can buy food' (interviewed on 27 February 2022). Irrespective of mitigative potential, the reconfiguration of labour geographies by solar power has left locals without land or livelihoods. As a result of solar enclosures, women are also forced to travel further distances in search of scarcer firewood and non-timber forest products (such as shea nuts), considerably increasing their labour burdens. According to a Christian Dagaaba peasant from Lawra, 'Now women have to go far to get these products, which affect productive hours and their livelihood' (interviewed on 5 February 2022). In sum, energy and livelihood injustices and dispossessions wrought by solar power ironically exacerbate economic and energy poverty in Kaleo and Lawra. Albeit bestowed with excellent solar irradiance, peasant vulnerability is overdetermined in the region imperilled by manifold social and climatic crises and manipulated by Northern paternalistic intervention in renewable energy transitions.

Conclusion

Energy geopolitics of the Global North are increasingly characterised by ensuring neoliberal development trajectories through renewable infrastructures in nations disproportionately vulnerable to (yet not responsible for) the climate crisis. The case of the Kaleo Lawra Solar Plant in Ghana typifies these trends, as the project was elaborated by a Spanish firm (Elecnor SA) with expertise in energy infrastructures, in collaboration with a German engineering firm (Tractebel) and with funding from the German development bank, KfW. The solar plant was built in coordination with the Volta River Authority, Ghana's electrical utility that now operates the solar plant and was recently restructured by the IMF. The project can be situated within broader efforts by transnational state actors and multilateral institutions to influence Africa's energy transition, such as the G20 Compact with Africa and Germany's Marshall Plan with Africa. The project is also a key component of Ghana's efforts to meet the climate goals established in their nationally determined contributions to the Paris Agreement, as well as targeting Sustainable Development Goals around affordable and clean energy.

A rising sun in Africa, the geopolitical ecologies of Ghanaian solar development illuminate darker vistas. Paradoxically, the Kaleo Lawra Solar Plant worsens the gendered vulnerability of resource-dependent peasants. Land enclosures for the project have dispossessed women of marginalised ethnic groups of access to vital firewood and economic trees, exacerbating their economic and energy poverty. Further, local people were not provided employment at the solar plant, compounding livelihood losses from encircled farmlands. This solar plant exemplifies broader dispossessionary energy transitions in the Global South that sustain colonial-capitalist production relations and Northern hegemony under the auspices of saving a world-system in crisis. Unless the land and production relations in Kaleo and Lawra are transformed, the local gendered injustices wrought by energy geopolitics are cognate aspects of solar power in Ghana that jeopardise the actualisation of a just transition and lock in a long-term fight for light. The sad trends brought to light mark the revelatory power of a focus on energy geopolitics and gendered injustices as cognate aspects of solar energy transitions.

References

Adusah-Poku, Frank, and Takeuchi, Kenji. 2019. Energy poverty in Ghana: Any progress so far? *Renewable and Sustainable Energy Reviews* 112: 852–64.

Bezner Kerr, Rachel, Naess, Lars O., Allen-O'Neil, Bridget, Totin, Edmond, Nyantakyi-Frimpong, Hanson, Risvoll, Camilla, Rivera Ferre, Marta G., López-i-Gelats, Feliu, and Eriksen, Siri. 2022. Interplays between changing biophysical and social dynamics under climate change: Implications for limits to sustainable adaptation in food systems. *Global Change Biology* 28(11): 3580–604.

Bigger, Patrick, and Neimark, Benjamin D. 2017. Weaponizing nature: The geopolitical ecology of the US Navy's biofuel program. *Political Geography* 60: 13–22.

BMZ. 2021. The Marshall Plan with Africa—Review and outlook. *Federal Ministry for Economic Cooperation and Development, Government of Germany*. Available at: https://www.bmz .de/resource/blob/86828/3357dcbd9969cb774b6fdeb7dfd75861/marshall-plan-review -outlook-4-years-ba-data.pdf

Bradshaw, Michael J. 2009. The geopolitics of global energy security. *Geography Compass* 3(5): 1920–37.

Cantoni, Roberto, Caprotti, Federico, and de Groot, Jiska. 2022. Solar energy at the peri-urban frontier: An energy justice study of urban peripheries from Burkina Faso and South Africa. *Energy Research & Social Science* 94: 102884.

CwA. 2017. Investment Opportunities—Ghana. G20 Compact with Africa. Available at: https:// www.compactwithafrica.org/content/dam/Compact%20with%20Africa/Prospectus/G20 _Compact_with_AfricaGhana_v5.pdf

Dalby, Simon. 2014. Rethinking geopolitics: Climate security in the Anthropocene. *Global Policy* 5(1): 1–9.

Elecnor. 2023. Solar photovoltaic: Kaleo and Lawra. Available at: https://www.grupoelecnor.com /resources/files/1/projects/en/referencia-kaleo-y-lawra-ghana-en.pdf

Ghana Statistical Service. 2020. Multi-dimensional poverty. Ghana Statistical Service, Government of Ghana. Available at: https://statsghana.gov.gh/gssmain/fileUpload/pressrelease /Multidimensional%20Poverty%20Ghana_Report.pdf

Government of Ghana. 2015. Ghana's intended nationally determined contribution (INDC) and accompanying explanatory note. Ghana: Government of Ghana.

IMF. 2016. Ghana: Letter of intent, memorandum of economic and financial policies, and technical memorandum of understanding. *International Monetary Fund*. Available at: https://www.imf .org/external/np/loi/2016/gha/091616.pdf

IPCC. 2022. *Climate Change 2022: Impacts, Adaptation and Vulnerability. Contribution of Working Group II to the Sixth Assessment Report of the Intergovernmental Panel on Climate Change*, edited by H.-O. Pörtner, D.C. Roberts, M. Tignor, E.S. Poloczanska, K. Mintenbeck, A. Alegría, M. Craig, S. Langsdorf, S. Löschke, V. Möller, A. Okem, and B. Rama. Cambridge and New York: Cambridge University Press. https://doi.org/10.1017/9781009325844

MESTI. 2015. Ghana national climate change master plan—Action programmes for implementation: 2015–2020. *Ministry of Environment, Science, Technology and Innovation, Accra. Government of Ghana*. Available at: https://www.weadapt.org/sites/weadapt.org/files/2017/ghana _national_climate_change_master_plan_2015_2020.pdf

MESTI. 2021. Ghana: Updated Nationally Determined Contribution under the Paris Agreement (2020–2030). *Environmental Protection Agency, Ministry of Environment, Science, Technology and Innovation, Accra. Government of Ghana*. Available at: https://unfccc.int/sites/default/files /NDC/2022-06/Ghana%27s%20Updated%20Nationally%20Determined%20Contribution %20to%20the%20UNFCCC_2021.pdf

Ministry of Energy. 2022. Sector overview. *Ministry of Energy, Republic of Ghana*. Available at: https://www.energymin.gov.gh/sector-overview#:~:text=Ghana%20has%20since %20achieved%20an,yet%20to%20be%20fully%20exploited

Moore, Jason W. 2017. The Capitalocene, part I: On the nature and origins of our ecological crisis. *The Journal of Peasant Studies* 44(3): 594–630.

Nartey, Laud. 2022. Kaleo solar project: Most of the workers used for construction were locals from nearby towns — Akufo-Addo. *3 News*, 24 August. Available at: https://3news.com/kaleo-solar -project-most-of-the-workers-used-for-construction-were-locals-from-nearby-towns-akufo -addo/

Nuru, Jude T., Rhoades, Jason L., and Sovacool, Benjamin K. 2022. Virtue or vice? Solar micro-grids and the dualistic nature of low-carbon energy transitions in rural Ghana. *Energy Research & Social Science* 83: 102352.

Nyantakyi-Frimpong, Hanson. 2020. Unmasking difference: Intersectionality and smallholder farmers' vulnerability to climate extremes in Northern Ghana. *Gender, Place & Culture* 27(11): 1536–54.

Sareen, Sareen. 2022. Drivers of scalar biases: Environmental justice and the Portuguese solar photovoltaic rollout. *Environmental Justice* 15(2): 98–107.

Stock, Ryan, and Birkenholtz, Trevor. 2020. Photons vs. firewood: Female (dis)empowerment by solar power in India. *Gender, Place & Culture* 27(11): 1628–51.

Tractebel. 2022. Sun gives Ghana even more power. *Tractebel*, 15 December. Available at: https://tractebel-engie.com/en/news/2022/sun-gives-ghana-even-more-power

VRA. 2020a. Environmental Impact Assessment Report: Non-Technical Summary. 35 MW Solar Power Project (SPP): Upper West Regional Project Sites. Volta River Authority.

VRA. 2020b. Sustainability report 2020. Volta River Authority. Available at: https://www.vra.com/media/2021/VRA%20Sustainability%20Report%202020.pdf

VRA. 2022. Commissioning of the Kaleo solar power plant. Volta River Authority. Available at: https://www.vra.com/media/2022/kaleo-solar-power-plant.php

10
Governing solar supply chains for socio-ecological justice
Dustin Mulvaney

Intersections of solar deployment and governance

Solar energy is rapidly becoming a major source of electricity for human civilisation. As solar deployment rises to terawatts levels, the industries that supply solar energy production with specific materials, natural resources, labour and lands will reconfigure socio-ecological relations. The development of solar power commodity chains will drive increased demand for minerals and metals and produce new geographies of metallurgy, smelting, semiconductor and glass manufacturing, solar photovoltaic (PV) module assembly and landscapes of electricity generation. This raises several important questions because as solar production scales, it may reproduce existing inequalities, have unintended consequences, or cause otherwise easily mitigated impacts. This chapter argues that scaling the production and development of solar power devices and infrastructures with impact mitigation in mind can improve socio-ecological outcomes. It identifies innovative governance interventions aimed at mitigating increased pressures on global change caused by rapidly scaling up solar power production and challenges and opportunities to engage in effective impact mitigation strategies.

The activities that govern the production of solar electricity and the commodity chains that constitute it as a material object, are disparate activities in time and space. Unlike other industrial sectors, for example aluminium or steel, or coal, or oil and gas industries, there are no overarching entities or centres where sets of rules that might govern solar and ancillary industries are developed. Instead, the governance of solar energy is the result of a set of scattered policies that touch on various aspects of many different supply chains, manufacturing operations, similar or homologous products, siting rules and processes, and end-of-life management practices. This matters, because to suggest aspirational ideas about sustainability and environmental justice in this context means applying them in disparate multi-sited areas of manufacturing, zoning, economic investment, trade policy, private contracts and through whatever other tools, policy, or practices are available to shape solar production in ways that enhance socio-ecological justice.

Starting from the perspective that solar PV offers significant advantages to energy systems based on combustion, this chapter argues that solar PV could be made in ways that reflect principles of socio-ecological justice more thoroughly. Thus, socio-ecological justice constitutes a cognate aspect of solar energy transitions. Solar deployment across the world has already created instances where important environmental justice and ecological issues have arisen and not been resolved (Yenneti and Day, 2016). Labour health and safety, community impacts and land uses have each shown there are impacts across the life cycle in need of attention. Governing the solar industry beyond simply deployment, with attention to improved socio-ecological outcomes, could result in addressing these key impacts from solar development and production. For example, contemporary polysilicon manufacturing is significantly more carbon intensive than other products in the global economy owing to the very high proportion of coal heath and power and other geographic circumstances of where that production occurs, namely the most energy-intensive stages in coal-fired industrial parks. Utility-scale solar development has caused loss of quality farmland (Hoffacker et al., 2017) and critical ecological habitat (Hernandez et al., 2014) and adverse impacts on the landless and poor (Stock and Birkenholtz, 2020, 2021) and on the socio-cultural domain (Silva and Sareen, 2021). Problems that one might argue are important to resolve in the context of justice and sustainability are also relatively easy matters to mitigate or avoid through better public policy, standards, regulation or practice. Yet, there is very little progress pushing in these directions – very little to no advocacy, despite these efforts being arguably integral to the successful manifestation of a more just and sustainable industry.

Solar power commodity systems have few loci of power or sites of intervention to shape the broader system and coupled with a lack of environmental advocacy and engagement with labour issues and human rights, it seems challenging to work on these solutions in a comprehensive way to resolve some of the emergent contradictions. This makes it

Table 10.1 Multi-sited governance interventions currently shaping the geography and resource assemblage for photovoltaics production

Life cycle stage	Life cycle stage	Examples of policy interventions	Certifications, standards, guidance
Supply chain and manufacturing for solar but mostly other industries	Silica mining/processing	Environmental laws on mining, import restrictions, traceability requirements	Initiative for Responsible Mining Assurance, International Labor Standards on Occupational Safety and Health
	Metallurgical silicon	Air quality and labor laws, carbon capture requirements in coking	Global Electronics Council, Ultra Low Carbon Solar standard
Supply chain and manufacturing for solar industry	Polysilicon production	Import restrictions; border carbon adjustments	Global Electronics Council, Ultra Low Carbon Solar standard
	Wafer/cell manufacturing	Domestic content requirements, chemical stewardship laws	Energy Star, Scope 3 emissions accounting/climate action
	Module assembly	Low carbon materials or domestic procurement requirements	Cradle to Cradle Electronic Product Environmental Assessment Tool, Energy Star equipment
Land use and development	Siting and deployment	Land use restrictions, land cover reporting requirements, incentives for light on land approaches	Reports by the Center for Biological Diversity, The Nature Conservancy, Wilderness Society
End-of-life	Recycling and disposal	Electronic and electrical equipment waste circular economy policies, EU WEEE-RoHS Directives, Washington state in the US	Basil Action Network's e-Stewards, Solar Scorecard, Global Electronics Council's Electronic Product Environmental Assessment Tool

critically important to find those specific sites of governance that could be leveraged. Table 10.1 outlines several places where policy already intersects with governance and private certifications and standards, organised by the stages of production.

In what follows, this chapter will trace the stages of production that take natural resources and labour and turn it into solar electricity. Each section aims to identify important considerations for socio-ecological justice, highlight a few intervening factors that currently shape how and where solar power commodity systems are made, built and deployed, and offer suggestions for new opportunities to enhance the benefits that solar power can deliver. The goal is, like other contributions to this volume, to illuminate the scattered peripheries of activities – in this case in terms of the cognate aspect of socio-ecological justice – that constitute the solar energy we see made in the world today, so they can be augmented or leveraged for better solar energy governance at multiple sites along this global commodity chain.

Solar supply chains and manufacturing

Most depictions of the solar energy supply chain start with polysilicon, but really, making the semiconductors for PV begins with mining quartzite found at hard rock mines or gravel mines to produce a silicon metal, or metallurgical grade silicon. This material is further refined eventually to polysilicon, where it is melted, cast and cut into ingots that are 99.999999 per cent silicon. Most of a solar panel by weight is glass, another key component derived from another kind of quartz. As with most mining activities, quartzite mining impacts can be a source of land use conflict, ecosystem degradation, water overuse, or non-point effluents. The largest quartzite mines are in Quebec and Labrador, Canada, with historical mines in the US in North Carolina and Alabama. Other regions contribute too, but impurities can make this resource limited to specific geographies. Key occupational health issues include handling and processing of quartz, which can expose workers to silicon dust, potentially leading to silicosis, one of the oldest occupational diseases in human civilization (Poinen-Rughooputh et al., 2016). Most of the quartz mined and processed at this stage is made into other products for other industries, such as the steel industry that turns it into ferrosilicon used as a de-oxidizer in the production process. The solar industry is one of many buyers of quartz for its semiconductors, glass and other quartz-derived products, such as quartz crucibles for polysilicon manufacturing, which are nearly all produced at a single high purity quartz mine in North Carolina.

Quartz is next processed into metallurgical silicon, or silicon metal. This industrial metal also finds its way into multiple sectors, including aluminium alloys, silanes and silicones; somewhere on the order of 15 per cent of silicon metal goes to the solar industry. The facilities can have impacts to air quality and the process directly produces carbon dioxide to remove the oxygen from quartz. These facilities are often at risk for explosions and fires from silica dust, posing a threat to worker safety, especially if they are poorly maintained (Habashi, 2012).

Silicon metal is made into polysilicon, which can involve both dangerous and laborious work. In 2020, several explosions at polysilicon manufacturing facilities in Xinjiang, a region in northwest China, within weeks of each other took down 20 per cent of global polysilicon production, resulting in manufacturing delays and steep price increases in the spot market for polysilicon. Explosive hydrogen, silicon dust, refrigerants, and other heat transfer fluids require care and special training in handling to avoid fires and explosions. There are also effluents of hazardous silicon tetrachloride wastes and corrosive hydrochloric that need to be treated with best practices in chemical stewardship. After production, polysilicon crystals are broken into chunks before the next step in processing, and much of this repetitive work is done manually because of contamination concerns with mechanisation. Such dangerous and tedious work can be an environmental injustice in the workplace.

Silicon ingots are cast from polysilicon and cut into wafers which are then made into the solar cells that comprise a final silicon photovoltaic module. Solar cell manufacturing uses bulk and specialty chemicals as well as strong acids and bases, so environmental justice here means the industry needs to be a good neighbour with regard to emissions, effluents and accidents. A hydrofluoric acid spill at a solar cell manufacturing in China in 2011, caused by Jinko Solar, resulted in a fish kill and injured domestic farm animals who were bathed in or drank the contaminated water. This led to a protest where villagers stormed the facility and broke windows and equipment (Mulvaney, 2014).

Other key components in a typical PV supply chain include aluminium, silver, copper and thermoplastics. One class of technologies to generate solar power, not described here, are cadmium telluride thin films. Most thin films today use cadmium compounds to make a PV semiconductor (also with tellurium and/or sulfur). Since cadmium is a severely toxic metal in elemental form, this raises concern all along the supply chain back to smelters and zinc mining where it is often produced.

There has been little work to date on the issue of improved working conditions in solar manufacturing across its life cycle, with one major exception. Silicon metal and polysilicon production has recently been linked to human rights abuses in the Xinjiang autonomous region in northwestern China, where 45 per cent of the global supply of polysilicon is produced. The accusations assert that ethnic minority Uyghur, Kyrgyz, Kazakh, and Tibetan peoples are used in forced labour schemes for the production of silicon metal and polysilicon (Murphy and Elimä, 2021). These allegations are denied by the Chinese government and companies operating there, but they implicate the four largest manufacturers in the world. The charges assert that ethnic minorities are being transported to the region for retraining and re-education in labour camps. Visibility into what is happening on the ground is severely hampered as human rights observers have left the region, citing a lack of meaningful access to production facilities and workers, rendering standards and certifications that rely on factory audits ineffective.

The US government now restricts imports from places accused of forced labour through executive action and a law passed by Congress. Since 2021, Customs and Border Protection have detained imports of PV, some made by the largest solar manufacturers in the world. The Department of Homeland Security, the US agency that enforces laws related to customs and imports, has announced that subsidiaries of polysilicon makers Daqo, GCL, East Hope, and Hoshine would all be barred from selling products. The four largest solar panel manufacturers in the world all source from one or more of these suppliers. At least one of the polysilicon and silicon metal companies tied to forced labour were also associated with the Xinjiang Production and Construction Corps, which the US government considers a paramilitary organisation. Most recently, Customs and Border Protection began detaining PV modules that cannot document their supply chain for quartzite, silicon metal and polysilicon. These imports were destined for polysilicon manufacturing facilities in the US and seized to enforce the Uyghur Forced Labor Prevention Act. In addition to 45 per cent of global production, Xinjiang is also the source for much of the spot market, where manufacturers are able to acquire additional polysilicon to blend in with their previously contracted source. Some estimate this means up to 90 per cent of the global supply of PV could be tainted by forced labour. This points to how deeply forced labour is embedded in the key materials used in solar production and difficulty in governing with anything but an import restriction without robust traceability rules.

There are a few voluntary standards under development that address some of the environmental, health, and safety issues in industry, energy efficiency and low carbon manufacturing practices, and avoidance of prison or forced labour such as the Green Electronics Council's (GEC)

Sustainability Leadership Standard and Ultra Low Carbon Standard criteria for solar (Global Electronics Council, 2023). These allow buyers to procure 'sustainability certified' or 'low carbon' PV from a registry for products that meet the standard. Similarly, the Silicon Valley Toxics Coalitions' Solar Scorecard aims to highlight companies that avoid prison labour and utilise the best in occupational health and safety at their facilities. Given the private and voluntary nature of these certifications, these efforts are limited to institutional buyers and private contracts that may stipulate purchasing off these registries or scorecards in private contracts.

Solar deployment

Utility-scale solar deployments raise land use questions because collecting this more diffuse source of power (than say fossil fuels) requires space (Smil, 1984). Energy sprawl from utility-scale solar projects is a leading cause of land use conflict across the American West (Mulvaney, 2017). These projects can be very large facilities with footprints on the order of square miles. Some solar projects have damaged cultural resources – geoglyphs, cremation sites and spiritually important locations. Others have resulted in significant local ecological impacts to habitats of important species. There are varied impacts from large utility-scale projects, and many of the worst impacts can be avoided by taking advantage of land with existing industrial uses. An example is the Western Solar Plan adopted by the US Bureau of Land Management, which attempts to direct projects onto previously disturbed or lower conflict sites, avoiding the sprawl onto 'greenfield' or otherwise undeveloped sites.

Opportunities to avoid habitat degradation through better land use selection are widespread. In addition to incentives to increase distributed energy resources in the built environment, there are several other solutions advocated by researchers and practitioners that could offer fewer land use conflicts over utility-scale solar power. Many of these projects offer multiple benefits in addition to solar power: water savings, agro-ecosystem enhancement, more productive land uses, and carbon sequestration – in addition to avoided habitat loss. Contaminated agricultural lands, abandoned landfills and mines, brownfields and other previously disturbed sites offer enough parcels to supply enough power to the US many times over (Hernandez et al., 2019). Integrating solar projects into existing and proposed infrastructures cost-effectively is a major challenge on its own, but land-sparing projects show potential for an enormous supply of renewable energy with minimal impacts to ecosystems and cultural resources.

Integrating PV into agriculture can help avoid land use change as well. US agriculture covers about 1.4 million square miles, with 87.6 million acres of soy beans and 92.1 million acres of corn crop in 2021 (USDA, 2022) – the latter sending 40 per cent to ethanol production. Imagine instead of supplying a transportation fuel (ethanol is generally 10 per cent of gasoline in the US), some small fraction of these corn and soybean lands were converted to solar farms producing energy for electric vehicles. US corn and soybean fields together could produce 25 terawatts of solar power – more power than used by humans today. Only a fraction would be needed, but solar projects sited on former farmlands and developed with an emphasis on ecological restoration could yield multiple benefits. Recent research finds grassland restoration at solar farms in the Midwestern US (mostly lands formerly dedicated to corn/soybean row crops) increases pollinators, sequesters carbon, reduces sediment loss and retains water in soils (Walston et al., 2021).

Agrivoltaics – typically agricultural row crops interspersed with PV – can coproduce energy and crops from the same land. Rows of crops and solar panels contribute side by side, yield less than either monocultures or solar would alone in a field, but more in total, especially with crop quality improvements or reduced input costs (Chae et al., 2022). Partial shading from agrivoltaics can provide reduced water stress, water savings and offer improved crop quality. Water can be the limiting factor for some crops, and the solar arrays' shade can make microclimates that improve soil moisture. Crops that are already grown in hoop houses or greenhouses, like berries, make prime candidates.

Floating solar deployments are opportunities for multiple uses of space over water, including aquaculture, mariculture or integration with other coastal or energy infrastructures or reservoirs for wastewater treatment. But floating solar PV may also encounter many use conflicts, and water bodies are used for navigation and flood control. In May 2022, China's Ministry of Water Resources banned wind and solar floatovoltaics (a play on floating 'voltaics') from freshwater bodies to protect the hydrological integrity of water bodies and flood management (S&P Global, 2022). They claimed solar and wind projects can obstruct the steady flow of water and damage river banks and dykes, which are key to protecting communities and agriculture, resulting in dismantling of the 1 GW/$1.2 billion Tiangang Lake floating solar project in Jiangsu province (China Dialogue, 2022). The social and environmental impacts of these infrastructures need to be assessed to make sure projects are done in ways that do not compromise flood control or contribute to erosion. Moreover, they require study and best practices to ensure they do not result in unintended negative impacts to birds or to aquatic and benthic species.

Land use governance for solar PV has been limited to a few spatial planning processes, such as the Desert Renewable Energy Conservation Plan and Western Solar Plan in the US. The state of Minnesota has groundcover rules for solar that require reporting vegetation change and require the reseeding of pollinators. But absent direct ways to regulate land use, say for example prohibiting blade and roll techniques on sensitive soils, best land use practices for now need to be a part of the contracts between project developers and offtakers.

Solar electricity's end-of-life

The end-of-life of electronics products conjures up images of artisanal workers using rudimentary tools to recover valuable materials embedded in a matrix of toxins. Photovoltaics raise some similar concerns due to the recipe of a toxic material – in most cases lead, but with some thin films, its cadmium – firmly embedded with the valuable materials (chiefly lead in crystalline silicon, and cadmium compounds in the CdTe thin films, copper, and glass).

Circular economy priorities for photovoltaics in the energy transition are best leveraged by policies that encourage product stewardship, such as extended producer responsibility (EPR), a form of the 'polluter pays' principle but for end-of-life products. Europe's Waste and Electrical and Electronic Equipment Directive has helped encourage the collection of end-of-life PV panels in Europe, where over 90 per cent of modules are recycled. In the US, the lack of electrical equipment management policy and emphasis only on hazardous waste has led to less than 5 per cent of PV modules being recycled. Places considering EPR policies still face pushback on the question of who pays, as often the state has stepped into the support programmes in lieu of state support, which is less with the intention of EPR policy. Finally, an e-stewards recycling standard for PV developed by the Basel Action Network that certifies facilities that set up and train workers on proper handing protocols can help ensure best practices to avoid occupational injury and exposures.

For places outside Europe, recycling governance opportunities exist solely in the spaces of private governance and certifications. The aforementioned GEC sustainability standard offers points for end-of-life management of modules, although it is not required to get the certification. A global commitment to govern flows of end-of-life PV to safe and responsible disposal and recycling will help facilitate a circular economy in these materials, driving innovations in closed-loop recycling, innovations in glass recovery and modules with longer expected lifetimes (Mirletz et al., 2022).

Governing solar towards just socio-ecological relations

As a nascent industry, solar PV still can get the land, labour and end-of-life questions right, but it will require convincing politicians, environmental organisations and the public that these alternatives are viable and worthy of resources. It is a false choice to frame conservation, quality jobs, or chemical and product stewardship in opposition to renewable energy when other alternatives are available. Some of these efforts require further research and development, but researchers have already identified and characterised many multifunctional and synergistic deployment strategies. The design and implementation of policy and procurement incentives from buyers of renewable energy might be seen as opportunities to bring more elements of sustainability and justice into consideration. In other words, private voluntary standards, despite their shortcomings and the limits of ecolabels in general, could provide meaningful direction in shaping solar supply chains away from some of the short-term challenges, namely those associated with accusations of forced labour and high carbon intensity. This is not an endorsement of private regulations per se, but a reflection of this moment where the advocates that typically look at questions of justice and sustainability are instead worried that engagement with these issues will undermine solar deployment more broadly.

Three overarching themes of labour, land and end-of-life can be drawn together for a framework for better solar production and applications. They constitute important interlinked elements of socio-ecological justice as a cognate aspect of solar energy transitions. But the lack of a central governing authority, and competition against incumbent mostly fossil energy, make it challenging to gain momentum to find champions for more discretion and selectivity when it comes to improving the impacts of solar production, whether that is product stewardship, land use, environmental health and safety, or safe and responsible product disposal. Without a focused and targeted campaign, efforts to govern solar energy towards better, more socio-ecologically just outcomes will remain limited.

References

Chae, Seung-Hun, Kim, Hye J., Moon, Hyeon-Woo, Kim, Yoon H., and Ku, Kang-Mo. 2022. Agrivoltaic systems enhance farmers' profits through broccoli visual quality and electricity production without dramatic changes in yield, antioxidant capacity, and glucosinolates. *Agronomy* 12(6): 1415.

China Dialogue. 2022. Wind and solar projects banned from freshwater bodies. *China Dialogue*, 1 June. Available at: https://chinadialogue.net/en/digest/wind-and-solar-projects-banned-from-freshwater-bodies/

Global Electronics Council. 2023. GEC announces initiative to advance decarbonization of solar panel production. Available at: https://globalelectronicscouncil.org/news/gec-announces-initiative-to-advance-decarbonization-of-solar-panel-production/

Habashi, Fathi. 2012. A review: Pollution problems of the metallurgical industry. *Revista del Instituto de Investigación (RIIGEO), FIGMMG-UNMSM* 15(29): 49–60. Available at: https://revistasinvestigacion.unmsm.edu.pe/index.php/iigeo/article/view/2204/1915

Hernandez, Rebecca R., Armstrong, Alona, Burney, Jennifer, Ryan, Greer, Moore-O'Leary, Kara, Diédhiou, Ibrahima, Grodsky, Steven M., Saul-Gershenz, L., Davis, R., Macknick, J., and Mulvaney, Dustin. 2019. Techno–ecological synergies of solar energy for global sustainability. *Nature Sustainability* 2(7): 560–68.

Hernandez, Rebecca R., Easter, S., Murphy-Mariscal, Michelle L., Maestre, Fernando T., Tavassoli, M., Allen, Edith B., Barrows, Cameron W., Belnap, J., Ochoa-Hueso, R., and Ravi, S. 2014. Environmental impacts of utility-scale solar energy. *Renewable and Sustainable Energy Reviews* 29: 766–79.

Hoffacker, Madison K., Allen, Michael F., and Hernandez, Rebecca R. 2017. Land-sparing opportunities for solar energy development in agricultural landscapes: A case study of the Great Central Valley, CA, United States. *Environmental Science & Technology* 51(24): 14472–82.

Mirletz, Heather, Ovaitt, Silvana, Sridhar, Seetharaman, and Barnes, Teresa M. 2022. Circular economy priorities for photovoltaics in the energy transition. *Plos One* 17(9): e0274351.

Mulvaney, Dustin. 2014. Solar's green dilemma. *IEEE Spectrum* 51(9): 30–33.

Mulvaney, Dustin. 2017. Identifying the roots of Green Civil War over utility-scale solar energy projects on public lands across the American Southwest. *Journal of Land Use Science* 12(6): 493–515.

Murphy, Laura, and Elimä, Nyrola. 2021. *In Broad Daylight: Uyghur forced labour in global solar supply chains*. Sheffield, UK: Sheffield Hallam University Helena Kennedy Centre for International Justice. Available at: https://shura.shu.ac.uk/29640/1/Murphy-InBroadDaylight%28VoR%29.pdf

Poinen-Rughooputh, Satiavani, Rughooputh, Mahesh S., Guo, Yanjun, Rong, Yi, and Chen, Weihong. 2016. Occupational exposure to silica dust and risk of lung cancer: An updated meta-analysis of epidemiological studies. *BMC Public Health* 16(1): 1–17.

S&P Global. 2022. China restricts solar, wind power projects in inland waters, cites flood control. *S&P Global,* 9 June. Available at: https://cleanenergynews.ihsmarkit.com/research-analysis/china-restricts-solar-wind-power-projects-in-inland-waters-cit.html

Silva, Luís, and Sareen, Siddharth. 2021. Solar photovoltaic energy infrastructures, land use and sociocultural context in Portugal. *Local Environment* 26(3): 347–63.

Smil, Vaclav. 1984. Views: On energy and land: Switching from fossil fuels to renewable energy will change our patterns of land use. *American Scientist* 72(1): 5–21.

Stock, Ryan, and Birkenholtz, Trevor. 2020. Photons vs. firewood: Female (dis)empowerment by solar power in India. *Gender, Place & Culture* 27(11): 1628–51.

Stock, Ryan, and Birkenholtz, Trevor. 2021. The sun and the scythe: Energy dispossessions and the agrarian question of labor in solar parks. *The Journal of Peasant Studies* 48(5): 984–1007.

USDA. 2022. U.S. crop acreage. Available at: https://www.fsa.usda.gov/news-room/efoia/electronic-reading-room/frequently-requested-information/crop-acreage-data/index

Walston, Leroy J., Li, Yudi, Hartmann, Heidi M., Macknick, Jordan, Hanson, Aaron, Nootenboom, Chris, Lonsdorf, Eric, and Hellmann, Jessica. 2021. Modeling the ecosystem services of native vegetation management practices at solar energy facilities in the Midwestern United States. *Ecosystem Services* 47: 101227.

Yenneti, Komali, and Day, Rosie. 2016. Distributional justice in solar energy implementation in India: The case of Charanka solar park. *Journal of Rural Studies* 46: 35–46.

Index

Page numbers in *italics* are figures; tables are in **bold**.

aesthetics 8–11, 94–5
age 14
agrivoltaics 132
alternative energy solutions 57–65

Baker, Lucy 24, 33
Bamford, Anne 69
Barthes, Roland 70
Benin 59–60, 61–3
Bihar (India) 59, 61, 62, 65
Bressand, Albert 68

cadmium 129, 133
California 98
carbon capture and utilisation and storage
    (CCUS) technologies 71, 76–9
Celtic Freeport plan 67, 76
China 130, 132–3
Christophers, Brett 25
citizen energy communities (CECs) 104–5,
    105, 106, 109, 110
Clean Energy Package 36
Cleanwatts (Portugal) 109
climate emergency, and Spain 39–42
climate mitigation 1, 18, 20, 118
cognate aspects, term defined 1–2
commodity chains 125–6, 128
Community Empowerment through Solar
    Energy (Mexico) 49
community solar projects
    Portugal 97–111, *100*, *102*, *105*
    see also energy communities
cooperatives 100
Coopérnico (Lisbon) 101–11, *102*
Cotonou (Benin) *see* Benin
COVID-19
    Mexico 51
    Spain 39, 42

deregulation 26–7
de-risking 26–7
developers 8, 27, 28–9, 31–3, 91, 133
direct pay 31
dispossessions, land/resource 4, 19, 119, 121

eco-communities 37
Ekins, Paul 68
electricity capital 24, 25
    tax equity partnerships 29–32, *30*
emotions
    and acceptance of renewable energy 68–9,
    70

Mexican case study 52–4
empowerment 18, 24, 49, 51, 52, 55
end-of-life electronics 133–4
energy access 13
energy communities 12–16
    emotional 54
    Spain 35–44
energy democracy 4, 24, 33
energy geopolitics 18–19, 20, 115
    see also Kaleo Lawra Solar Plant (Ghana)
energy peripheries 2
energy poverty
    Mexico 49, 50, 53
    Spain 39, 42, 43
ensconcement
    Jaipur and Lisbon 81, 93–5
        contrasting situated visual
            ethnographies 83–92, *84*, *86–92*
        urban geographies 82–3
environmental nongovernmental organisations
    (ENGOs) 102
European Commission's Renewable Energy
    Directive 36, 104–6
European Union Directives 38, 104, 133
exclusion 2, 49
extended producer responsibility (EPR) 133

finance 16, 17–18
    US 23–33
financialisation 25, 33
floating solar deployments 132–3
fuel pumps 81, 85, *86*, 93

games, training 40–1
gender 14, 18
    and energy injustices in Ghana 115–22,
    116
geopolitical ecology 117–20
Germany, partnership with Ghana 118–19
Ghana, gendered energy injustices 115–22, *116*
Global South
    urbanisation 57–65
    see also Ghana
governance 37–9, 125–34, **127**
Green Electronics Council (GEC) standards
    130–1, 133–4

Hoplass Solar farm (South Wales) 73–6, *75*, 78
human rights 126, 130

India *see* Bihar
Industrial Decarbonisation Research and
    Innovation Centre (IDRIC) 67

inequalities 13, 16, 110, 125
    and energy communities 37, 43
    Ghana 117
    Global South 63
    urbanisation 83
Inflation Reduction Act (IRA) (2022) (US) 23,
    24, 31
injustices, gender 18–19, 115–22, *116*
investment tax credits, US 23–33, *30*, 97

Jaipur (Rajasthan, India) 81, 93–5
    contrasting situated visual ethnographies
        with Lisbon 83–92, *84*, *86–8*
    urban geographies 82–3
Jorge, Nuno Brito 101
justice
    socio-ecological 19–20, 125–34
    see also injustices

Kaleo Lawra Solar Plant (Ghana) 116–18, *116*,
    119
    and gendered solar energy injustices
        120–2

land use 5, 14, 134
    and aesthetics 8
    conflict 2, 3, 20
        Ghana 119, 121
        and supply chains 126, **127**, 131–3
        US 131–2
    dispossession 4, 19, 117, 119, 121
    governance 133
Lebanon 60, 62–3, 65
LIGHTNESS project 41–2
Lisbon 81, 93–5
    contrasting situated visual ethnographies
        with Jaipur 83–92, *89–92*
    Coopérnico 101–11, *102*
    urban geographies 82–3

ManzaEnergía 42
Menlo Park laboratories 26
Mexico, social situatedness 47–55
mini-grids 60, 63–5
Morgan, J.P. 26

net-metering 35, 87, 108
Network of Solar Communities (Mexico) 49

offtakers 29, 133
oil refineries 73–6, *75*

partnership flips 30
Pembroke Valero oil refinery (South Wales)
    73–6, *75*, 78
policy **127**, 128–31, 134
    European Union Directives 38, 104
politics 16, 21, 59
    see also energy geopolitics
polysilicon 126, **127**, 128–30
Port Talbot Dock development (South Wales)
    76–8, *77*
Portugal, community solar 97–111, *100*, *102*,
    105
prosumers 10, 37, 40, 41, 44, 104–5

Public Utility Regulatory Policies Act of (1978)
    27
pumpsets *87*

quartzite 128, 130

Rajasthan see Jaipur
renewable energy certificates (RECs) 29
renewable energy communities (RECs) 104–6,
    105, 107–9
renewable energy cooperatives 100
renewable energy research organisations 48–9
RESCHOOL project 40
ResCoop 101, 103, 104, 109
Roosevelt, Franklin D. 26

scalar cognates 11–16
self-consumption systems 104, 105–8
silicon metal 129, 130
situated knowledge 47
small-scale energy production 8
smart cities 81, 82
Social Demand of Energy 50
social innovation, and energy communities
    37–9, 43
social movements 98, 99
    see also Coopérnico (Lisbon)
social situatedness 13–14, 16
    Mexico 47–55
    see also visions/visual media
socio-ecological justice 19–20, 125–34
solar development process, US 27–9
solar farms
    portrayal in visual media 67–9, 78–9
        Pembroke Valero oil refinery/Hoplass
            Solar Farm 73–6, *75*
        reading solar landscapes 71
        solar-superplace acceptance/CCUS
            76–8, *77*
        solar/wind farm photograph 72–3, *73*
        visual analysis/psychosocial
            interpretation 70–1
solar support schemes 17
solar thermal 48, 49, 50, 51, 54, 88, *89*
solar trees 91–2, *91*
South Wales Industrial Cluster (SWIC) 67, 71
Spain, energy communities 35–44
SuperPlaces 76–8, *77*
supply chains 19, 125–34, **127**

tax equity partnerships 29–32, *30*

United Kingdom see Wales
United Nations Sustainable Development Goals
    38, 117
United States of America (US)
    finance 23–33, *30*
    policy 129, 133
    quartzite 128
urbanisation
    Global South 57–65
    see also Jaipur; Lisbon

van Leeuwen, Theo 70
visions/visual media

Wales 66–7
        Pembroke Valero oil refinery/Hoplass
            Solar farm 73–6, *75*
        and public acceptance 68–9
        reading solar landscapes 71
        solar farm/wind farm photo 72–3, *73*
        visual analysis/psychosocial
            interpretation 70–1
visual literacy 69

wellbeing 39, 50, 72
wind farms, South Wales 72–3, *73*
working conditions 129–30, 134
World Health Organization 39

zero-energy communities 37